Interrogating Eco-Literature and Sustainable Development

This book examines the issues of ecological crisis and sustainable development through critical reading of literary texts. By analyzing writings of Rabindranath Tagore, Amitav Ghosh, Gerard Manley Hopkins, Hannah Arendt, and Lawrence Buell, it discusses themes like oriental representations of ecological consciousness, environmental evocations, misogyny and its postmodern creations, tracing nature's footprints in English literature, statelessness and consequent environmental refugees, ecocriticism and comics, and absolute trust in the goodness of the earth.

The volume argues that within the ambit of debates between ecological threats and socio-economic concerns, culture plays a vital role, particularly in relation to parameters such as identity and engagement, memory and projection, gender and generations, inquiry and learning, well-being and health.

This book will be of interest to scholars and researchers of cultural studies, English literature, social anthropology, gender studies, sustainable development, environmental studies, ecological studies, development studies, and postcolonial studies.

Sharbani Banerjee Mukherjee is Associate Professor of English in the Postgraduate Department of English at Trivenidevi Bhalotia College, Raniganj, West Bengal, India. An alumnus of Visva Bharati, Santiniketan, her PhD thesis was on Lawrence Durrell's *The Alexandria Quartet*. Her areas of interest include American literature, post-fifties British literature, Postcolonial Studies, Partition literature, Ecocriticism and Gender Studies. She has co-edited *The American Novel from Hawthorne to Heller: Cultural Contexts and Critical Perspectives* (2019). She is Associate Editor of *Literary Oracle*, journal of the Department of English, Berhampur University, Odisha.

Soumitra Roy is Associate Professor of English at Kazi Nazrul Islam Mahavidyalaya, Asansol, West Bengal, India, and has a teaching experience of over 21 years. He is an alumnus of Burdwan University, West Bengal. His areas of interest include postcolonial studies, film studies, and sports literature. His publications include "Consciousness of a Modern Indian Nation: Analysing Gender Violence from New Delhi to Kamduni". He is also a reviewer of *Literary Oracle*.

Interrogating Eco-Literature and Sustainable Development

Theory, Text, and Practice

Edited by Sharbani Banerjee Mukherjee and Soumitra Roy

Routledge
Taylor & Francis Group
LONDON AND NEW YORK

First published 2023
by Routledge
4 Park Square, Milton Park, Abingdon, Oxon OX14 4RN

and by Routledge
605 Third Avenue, New York, NY 10158

Routledge is an imprint of the Taylor & Francis Group, an informa business

British Library Cataloguing-in-Publication Data
A catalogue record for this book is available from the British Library

Library of Congress Cataloging-in-Publication Data
Names: Mukherjee, Sharbani Banerjee, editor. | Roy, Soumitra (College teacher), editor.
Title: Interrogating eco-literature and sustainable development: theory, text and practice / edited by Sharbani Banerjee Mukherjee and Soumitra Roy.
Description: Abingdon, Oxon; New York: Routledge, 2023. | Includes bibliographical references. | Summary – Provided by publisher.
Identifiers: LCCN 2022055067 (print) | LCCN 2022055068 (ebook) | ISBN 9781032206653 (hardback) | ISBN 9781032469614 (paperback) | ISBN 9781003383970 (ebook)
Subjects: LCSH: Ecocriticism. | Climatic changes in literature. | Ecology in literature. | Nature conservation in literature. | Conservation of natural resources in literature. | Ecoliterature–History and criticism.
Classification: LCC PN98.E36 I55 2023 (print) | LCC PN98.E36 (ebook) | DDC 809/.9336–dc23/eng/20230208
LC record available at https://lccn.loc.gov/2022055067
LC ebook record available at https://lccn.loc.gov/2022055068

ISBN: 978-1-032-20665-3 (hbk)
ISBN: 978-1-032-46961-4 (pbk)
ISBN: 978-1-003-38397-0 (ebk)

DOI: 10.4324/9781003383970

Typeset in Sabon
by Deanta Global Publishing Services, Chennai, India

Contents

Figures

Contributors

Chandan Bandyopadhyay is a gold medallist in MA in Economics from the University of Burdwan. He was a research fellow of UGC and obtained his PhD from the University of Burdwan in 2007. He joined Kazi Nazrul Islam Mahavidyalaya, Churulia, Asansol, West Bengal, as a Lecturer in Economics in 2002. Currently, he is serving as Associate Professor in Economics at Asansol Girls' College. He completed a UGC-funded Minor Research Project and has also edited two volumes on *Food Security in India and on Women's Economic Empowerment and Inclusive Growth.* He has research interest in gender economics, environmental economics, developmental economics, etc. He has published several research papers and has written chapters in edited books. Besides, he writes articles on various issues in Bengali newspapers quite frequently.

Abhisek Bhattacharya, MA (2007), PhD (2014), has been serving as Assistant Professor in English, Khandra College, West Bengal, since 2010. Specialized in Indian English literature and postcolonial studies, he takes keen interest in twentieth-century Indian poetry and contemporary speculations in Indian philosophy. His monograph titled *Life in Living: A Study of Poetry Written in English by Five Modern Bengali Poets* (1930–1980) is recently out in print from India and Michigan (Manak Publications, New Delhi).

Sajalkumar Bhattacharya is Professor and Head, Department of English, Kazi Nazrul University, Asansol, West Bengal. He wrote his PhD thesis on "Family-Nation Interface in Indian English Fiction with Special Reference to Select Novels of Salman Rushdie and Amitav Ghosh." He was awarded a UGC Fellowship for the research. He is also a recipient of the prestigious Charles Wallace Fellowship. His areas of interest include nineteenth-century British literature, Indian and other new literatures in English, Bangla literature, ecoterrorism, and biopolitics.

Saikat Chakraborty is an integrated MPhil PhD scholar in the Department of English, Kazi Nazrul University, Asansol. He has submitted his MPhil thesis titled "So Much in a Name!: Reading Genealogies of Vehicle Inscriptions." His recent publications include "Cthulhu and the Snake:

(Im)Possibility of Posthuman Ipseity" in international peer-reviewed journal *Consortium*. His paper titled "Breast and Its Surplus: Re reading of Mahasweta Devi's *Breast Giver* and Yogesh Pagare's *Mulakaram*" is forthcoming from Lexington books, an imprint of Rowman & Littlefield.

Nilanjana Chatterjee is Assistant Professor of English at Durgapur Government College, West Bengal. She completed her doctoral thesis titled "House as Space in the Fiction of Contemporary Women Writers of Indian Subcontinental Diaspora" in 2016 from the University of Burdwan. She has co-edited *Re-theorising the Indian Subcontinental Diaspora: Old and New Directions* (2020) for CSP and *Covid-19 in India, Disease, Health and Culture: Can Wellness Be Far Behind?* for Routledge (2022, forthcoming). She has authored *Reading Jhumpa Lahiri: Women, Domesticity and the Indian American Diaspora* (2022) for Routledge and book chapters titled "Women and Natural Resource Management in Naga Folktales and 'Peoplestories': Situating Easterine Kire's Fiction" (2022) for Rowman & Littlefield and "'Stand Aside Death ... Today Is My Day': Contextualizing the Naga Esotericism in Easterine Kire's Novels" (2022, forthcoming) for Routledge. https://orcid.org/0000-0003-0955-3480

Anindita Chatterjee is Associate Professor and Head of the Department of English, Durgapur Government College, West Bengal. She completed her PhD on "John Clare: Poetry of Madness" from Jadavpur University. She has completed two Minor Research Projects on *Representation of 'modern' India in the Fictions of Chetan Bhagat* (funded by UGC) and *Sari Story: Dressing the Indian Woman through History, 'Tracing the Origin, Growth and Evolution of the Traditional Indian Wear'* (funded by ICSSR, MHRD, GOI). She has co-edited *Re-theorising the Indian Subcontinental Diaspora: Old and New Directions* (CSP; London, 2020) and *Covid-19 in India, Disease, Health and Culture* (Routledge: London, 2022). She has published a book chapter in *Indian Feminist Ecocriticism* (Lexington Books, 2022) and articles in *Textile – Cloth and Culture* (Taylor and Francis) and *Contemporary Review of the Middle East* (SAGE). Her areas of interest include nineteenth-century British literature, gender studies, and popular culture. https://orcid.org/0000-0003-1244-5544

Sudipa Chowdhury, MA (1991), MPhil (1997), PhD (2015), is Associate Professor in Bengali, serving at Khandra College, West Bengal, for the past 25 years. Her area of expertise includes pre- and post-Independence Bengali fiction and historical novels in Bengali literature. She has presented papers at many state and national conferences.

Mohana Das is a creative writer, book reviewer, and a research scholar at Amity University, Kolkata. Her articles have been published in national and reputed journals. Her academic area seeks across multidisciplinary studies broadly addressing Indian writing style and forms, queer studies, African American narratives, modern poetry, gender studies experimenting on the tradition of writing of different time and space. Das is the author of *The Comprehensive Study of Tennessee Williams's The Glass Menagerie* with further publications in a row. She has been heading a talk show under iSPELL West Bengal and works as an Executive Committee member for them.

Shruti Das is Head, Department of English, and Director of the Centre for Canadian Studies at Berhampur University, Odisha. She has been a visiting lecturer in four universities in Poland. She was invited to present a paper in the Heads of Commonwealth Government Sponsored Conference 2015, in Malta. She has the distinction of being placed in the ALA Directory of Scholars, Princeton University, USA. She has won the prestigious ALSE Biennial Travel Award for 2019 for her paper on Indigenous tribes of Odisha. She has written the foreword for *Crisis of Imagination*, a book on ecocriticism, published by Peter Lang. She has published 11 books and over 70 research papers so far. She has been placed in the Bibliography of Commonwealth Writers for her first volume of poems. She is the editor of *Literary Oracle*, an interdisciplinary journal of literature.

Pinaki De is a graphic illustrator–designer who has designed almost 500 book covers to date for various renowned national and international publishers. He is also one of the editors of the prestigious annual magazine *Longform*. A TEDx speaker, he served as the Indian comics advisor of Mangasia, the biggest ever exhibition on Asian comics curated by Paul Gravett for the Barbican Centre, London. De juggles his creative work with academics as he is an Associate Professor of English at Raja Peary Mohan College, Uttarpara, West Bengal.

Sarbendu Bikash Dhar has a teaching experience of about 17 years. He was awarded the doctoral degree in Geography from the University of Calcutta in 2012. His areas of research interest include human migration and its social aspects; human development; environmental issues; application of RS and GIS in monitoring, solving, and management of various environmental problems. He has guided a number of PG dissertations and is now engaged in PhD supervision. Till date, he has published several research publications of national and international repute, book chapters, and a text book on environmental geography.

Sonam Jalan is a PhD research scholar at Bankura University, West Bengal, working on climate change in literature. She has pursued her master's and MPhil from Ravenshaw University, Cuttack, Odisha. She is currently engaged as a state-aided college teacher in the Department of English at Ramananda College, Bishnupur, West Bengal,. Her areas of interest include Anthropocene, climate change fiction, eco-films, refugee studies, and Indian English poetry.

Arunima Karmakar is Assistant Professor of English, Trivenidevi Bhalotia College, Raniganj. She obtained her master's in English from the University of Burdwan with specialization in literary theory and Old and Middle English literature. Karmakar is pursuing her PhD at Kazi Nazrul Mahavidyalaya, Asansol. She has special interest in travel writing and gender studies. She has presented several papers in various seminars and conferences.

Sujit Malick has been serving as Assistant Professor of English at Trivenidevi Bhalotia College, Raniganj, since 2015. Currently, he discharges the duty of a Coordinator at the Postgraduate Department of English. He has published many research articles in indexed journals and has also presented papers at national and international conferences in India and abroad. His areas of research interest are posthumanism and socio-linguistics.

Ashok K. Mohapatra is Professor of English at St. Xavier's University, Kolkata. He has taught at Sambalpur University and Kalinga Institute of Social Sciences, Odisha.

Mohapatra was a Fulbright scholar at Columbia and Yale Universities. He was a UGC visiting fellow at Jadavpur University, the University of Delhi, and the University of Burdwan.

Mohapatra has worked extensively in the fields of postcolonial and cultural studies. His latest publications are a co-edited volume titled *The American Fiction: From Hawthorne to Heller* (Macmillan, 2020) and an essay on the Holocaust poetry of Dan Pagis in *Journal of Trauma and Literary Studies* (University of Nebraska Press, 2021). He guest-edited an issue of *The Global South* for Indiana University Press and was an editor of *Sambalpur Journal of Literatures and Cultures*.

Sourav Kumar Nag is Assistant Professor and Head of the Department of English Literature and Culture Studies, Onda Thana Mahavidyalaya, Bankura University, West Bengal. He has published research articles and book chapters in several national and international journals and edited volumes. He is a member of the Association for Anglophone Postcolonial Studies (GAPS), Germany, Europe. He has been invited to deliver a lecture and chair a conference session at the University of Bremen, Germany, in the years 2019 and 2022. He has translated Abanindranath Tagore's *Nalak* to English.

Riman Rakshit is a PhD scholar in the domain of urban studies in the Department of English, Jadavpur University, West Bengal. She has been working on the Jewish communities in India for the past three years during which she has researched and written on the Israeli–Indian relationship. She is a Lecturer at Durgapur Women's College and the publisher of the government-registered monthly magazine *Kornofulir Dhare*. She has translated and published a varied range of poems, particularly from the

Middle East into Bengali. Her research interests include urban history, comparative religion, minority studies, and human geography.

Pradipta Sengupta is Associate Professor of English at MUC Women's College, Burdwan, West Bengal. He researched on Updike for both his doctoral dissertation ("The 'Hawthorne Novels' of John Updike") and postdoctoral research project ("Recasting Contemporary America: A Study of John Updike's 'Rabbit Tetralogy'"). His area of interest is American fiction, and apart from many national publications, he has published in such journals as *The Belgrade BELLS* and *The John Updike Review*, the project Knjizenstvo, and Lexington Book Series. He is a noted Updike scholar and has presented papers and chaired academic sessions at the national and international level.

Goutam Buddha Sural is Professor, Department of English, Bankura University, West Bengal, and has been teaching since 1990. He went to Bristol University, UK, in 2006 as a visiting fellow. He is the author of a book titled *Hopkins and Pre-Raphaelitism* and has edited four books on tribal life. He has presented papers in various conferences and seminars, and a number of his papers have been published in different national and international journals and books. His areas of interest are Victorian poetry, Indian English literature, and tribal and Dalit studies.

Preface

On a hot and sultry afternoon of August 2019, the faculty members of the Department of English of Trivenidevi Bhalotia College, Raniganj, and Kazi Nazrul Islam Mahavidyalaya, Churulia, got together with a proposal to host a national-level seminar, tentatively proposed for December of the same year. Among a host of topics discussed, the team zeroed in and finalized the title as "Re-thinking Literature: Eco-criticism and Socio-economic Concerns." The success of the seminar set about the process of publishing a book on the seminar proceedings with some of the papers that had been presented at the two-day ICSSR-sponsored national seminar.

The papers which had been read out presented a pretty large variety and thus the nomenclature of the upcoming book presented a sort of conundrum for the editors of the book. After much thought and taking into account the large variety and content of the papers presented, it was decided that the title *Interrogating Eco-Literature and Sustainable Development: Theory, Text, and Practice* would do adequate justice to the interdisciplinary approach that the seminar had aspired for. Needless to say, coming from the Department of English, most of the papers would of course have a literary bent, but with presenters belonging to Geography and Economics, the new title does justice to our multidisciplinary approach.

In 1987, the Brundtland Report for the first time mentioned the term *sustainable development*. The report is often taken to be the guiding principle of the environmental concerns that the twentieth-century human race is confronted with. However, to think that environmental concerns linked with modern development and a conscious debate on such matters is entirely post-modernist would be foolishness. Such debates and co-related ecocritical writings go as far back as 1662 with John Evelyn's *Sylvia* and in the writings of Hans Carl von Carlowitz, Frederick Augustus I, and Aldo Leopold. In 1962, Rachel Carson published her seminal and much-acclaimed book *Silent Spring* which changed human perspective on how we looked at the world around us. It brought together a multidisciplinary approach towards how we understood the environment upon which the very existence of the human race depended. Simultaneously it opened up a debate on how we should understand and balance the need for sustainable development and related socio-economic issues. Over the last 50 years, it has thus given to us a fresh field of study under a new nomenclature: ecocriticism.

Ecocritics firmly believe that human culture is related to the physical world and such criticism can be the answer for interlinking the life forms, their sustenance, socio-economic development, and thus expand the very notion of the 'world.' The definite link between nature and culture is often best represented by the literary treatment of such ecological and environmental issues, their textual representations, and so-called 'thematizations' of land/world and nature/human beings that influence the actions on the land. We may therefore often need to go back to Cheryll Glotfelty who suggested a three-tier development of ecocriticism. Within the ambit of the debate between ecological threats, socio-economic concerns, and ecocritical perspectives, culture plays a vital role, particularly concerning such parameters as identity and engagement, creativity and recreation, memory and projection, belief and ideas, gender and generations, enquiry and learning, well-being and health.

The topic thus addresses the concerns that environmentalists share today about the world that we live in and the socio-economic anxieties that bind human lives while looking at the various environmental issues. At the same time, it is hoped that the question of sustainable development shall be interrogated vis-à-vis environmental concerns arising across geographical territories. What then can be the role of literary texts when we look into the various challenges that lie ahead of mankind when Mother Earth's existence is under threat?

As we continue to talk on issues concerning the environment and its conservation, we find that there has been a tremendous effort on the part of the global community to try and protect Mother Earth, particularly concerning the issues of gas emissions and the greenhouse effect. The Paris Agreement of 2016 has taken a firm step forward in protecting Mother Earth. Yes, arguments might be forwarded that the Paris Agreement has its flaws but it is also true that issues related to environmental protection are an evolving process and no final word can ever be spoken on such matters. The Paris Agreement is an agreement within the United Nations Framework Convention on Climate Change (UNFCCC), dealing with greenhouse gas emissions, mitigations, and finance, signed in 2016. The agreement's language was negotiated by representatives of 196 state parties at the 21st Conference of Parties of the UNFCCC held at Le Bourget near Paris and was consensually agreed upon on 12 December 2015. As of December 2020, all 197 members of the UNFCCC have signed the agreement and 189 remain a party to it. Of the eight countries which are not a party to the law, the only significant emitters are Iran and the United States. The UNFCCC is an international environmental treaty addressing climate change, negotiated and signed by 154 states at the so-called Earth Summit held in Rio de Janeiro from 3–14 June 1992 and which environmentalists call the United Nations Conference on Environment and

Development (UNCED). The UNFCCC seeks for the stabilization of the greenhouse effect on the atmosphere at a level that would prevent dangerous anthropogenic human-induced interference with the earth's climate system. Such a level should be achieved within a timeframe sufficient to allow ecosystems to adapt naturally to climate change, to ensure that food production is not threatened, and to usher in sustainable economic development across the world and more particularly in the Third World countries.

This book presents to us an opportunity to look into the various aspects of the Sustainable Development Goals that have been enshrined as part of the United Nations Development Plan. The 17 SDGs have 169 targets and the member nations are striving to strike a balance in protecting the environment in their countries while promoting industrial and agricultural development to meet the basic needs of an ever-growing population. The balance is indeed very delicate and has its dynamism in achieving these SDGs in each country. Economic planning; budgetary allocation; and targets on health, poverty alleviation schemes, sanitation, and education all are important as are the issues of feeding innumerable mouths, providing employment opportunities, and creating healthy and safe shelters. Environmental issues thus often can take secondary or even tertiary priorities in this quest. Literary responses to such varied issues are an important index to what is being thought about globally regarding such variegated problems, issues, and their possible achievement of solutions in the context of sustainable development. The literary perceptions and deliberations have also highlighted that though ecocriticism is a fairly contemporary awareness, humanity has always been abundantly aware and appreciative of what nature offers and how it sustains us. Perceptions of natural bounty and its hermeneutic significance in human life have been expressed in all languages and at all times. But punitive measures to sustain the world and its ecological balance are the main concern today. We are hopeful that this book will help in understanding the enormity of the impending disaster and how sustainable development can slow that process.

For us, the editing of these chapters has been as much a matter of education as of sheer joy. Collaboration has brought out the best from both editors and in the backdrop of the recent pandemic that has ravaged our world over the past two years, a lot of work on this book has been done through remote access. We have learned that our digital world can also sustain our individual and collaborative world, can help us to develop skills, and make us contribute to the conservation of our environment by turning the entire editing of a book through a completely paperless process. Hopefully, this book will contribute to the research that is being done on sustainable development and environmental awareness around the world, reaching out to a global readership.

Sharbani Banerjee Mukherjee and Soumitra Roy
Raniganj, West Bengal

Acknowledgements

This book titled *Interrogating Eco-Literature and Sustainable Development: Theory, Text, and Practice* would not have been possible without the efforts of many people who have directly or indirectly, consistently, and intermittently helped us through constant encouragement, cajoling, guidance, and inspirations. We would first wish to thank Dr Asish Kumar Dey, Principal, Trivenidevi Bhalotia College, Raniganj, West Bengal, for his encouragement and leadership in first helping us to organize the ICSSR-sponsored national seminar titled "Re-thinking Literature: Eco-criticism and Socio-economic Concerns" and during editing of this volume. We also wish to thank Prof. A.K. Mallick, Teacher-in-Charge, Kazi Nazrul Islam Mahavidyalaya, Churulia, West Bengal, for his support.

We express our heartfelt gratitude to Dr Ashok K. Mohapatra, retired Professor of English, Sambalpur University, Odisha, and presently Professor of English, St Xavier's University, Kolkata, India, for consenting to write the introduction, which truly strikes the keynote and sets the deliberations in motion. We also thank and congratulate all the contributors for enriching this volume with their erudition and fresh thoughts.

We remain beholden to the authorities of Indian Council of Social Science Research, Eastern Region Centre, Kolkata, at the forefront of which has been the enterprise of its director, Dr Saibal Kar, and his team. Thanks are also due to our esteemed colleagues at the Department of English of the two colleges – Sujit Malick, Arunima Karmakar, Kasturi Joddar, Sumbul Nasim, and Pankaj Kumar Soren – for constantly keeping contact with the contributors of this edited volume and sifting through loads of email endlessly. We would also wish to thank the two state-approved college teachers – Souvik Dutta and Sayani Nayak – for their constant support.

We wish to thank Dr Nilanjana Chatterjee for helping us to dream big and then cross-checking most painstakingly almost every word of every chapter submitted. We wish to thank Dr Shruti Das for ushering us to Routledge India. We take this opportunity to express our gratitude to Saikat Chakraborty for going through the citations from time to time.

And finally, where would this edited volume be without Mr Shashank Shekhar, Ms Shoma Choudhury, Ms Antara Ray Chowdhury, Ms Brinda

Sen, and Ms Sunayina Dadhwal of Routledge India who have been in constant touch and have even tided over a pandemic to guide us through this volume.

Last but not the least, a big shout out for our respective spouses, Ananda Kumar Mukherjee and Anamika Roy, for their fortitude and patience as we forgot everything else apart from this book. Our children Nilanjana Mukherjee, Abhijeet Pattanaik, and Ayush Roy deserve special mention for appreciating our endeavour.

Dr Sharbani Banerjee MukherjeeSoumitra Roy

Introduction

Of Anthropocene: Far End of the Eco-critical Trajectory

Ashok K. Mohapatra

Hamlet's understanding of the eschatological scheme of "more things in heaven and earth" much larger and deeper than the philosophical disquisition at Wittenberg which he and Horatio attended, helps us, scholars from the disciplines of the Humanities, and particularly literary scholars, comprehend analogically how an uncanny and murky state of environmental affairs, imponderable to putative scientific reasoning, can originate from strange and remote causes. Like Hamlet, who can listen to the ghost, if not engage him in a dialogue, we embroil ourselves with the issues of environmental pollution already dead-locked among interest groups such as various environment protectionists, environment-sceptics, industries, corporate lobbyists, and various government and non-government policymaking bodies. The empirical truth claims and counterclaims fuelling rounds of unending debates, regardless of the fast-worsening environmental health of the earth, have bogged humans down to anxiety or confusion at best or indifference at worst. We witness all this and quietly register this scenario in our own ways, given the abiding concern of literature with humans in their natural and cultural milieus. Indeed, the Humanities seem to benefit considerably from the climate of suspicion that the objectivity and disinterestedness of the Natural Sciences and technology may often be a masquerade to push certain neoliberal economic agenda of development that is detrimental to the environment, biodiversity, and certain socio-economic forms, as well as functions that are directly dependent on the environment. The indispensability of human values and ethics as yardsticks for assessing the working of science and technology in this planet of ours makes the Humanities relevant.

Literature and the Social Sciences, both, build up their discourses on the subject of the environmental crisis affecting life on the earth, but literature has its distinct advantages over the Social Sciences, in that it is not stymied by empiricism, and it commands imagination to surpass the temporality of the present. It is also credited with an all-encompassing range of empathy, an uncanny power of intuition, and a keen analytical as well as interpretive power to tease meaning out of apparently disparate sets of ecological events that are spread over space and time in multiple schemes of concatenation, together with stray, contingent factors. Further, literary imagination,

DOI: 10.4324/9781003383970-1

following the Aristotelian principles of aesthetic probability and necessity, can conjure up the horrid vision of eco-dystopia and apocalypse looming large over the planet in a distant but foreseeable future. This kind of futuristic projection with cataclysmic changes in the environment affecting humans, plants, and animals – as in Stephen Baxter's *Flood* (2009) and his *Ark* (2009) or Margaret Atwood's *MaddAddam* (2013) – is almost beyond the ken of the Social Sciences. Even as the theme may be an ecological disaster on a large scale, the intensity and enormity of it are felt through the lives of individuals and families which literature is obliged to represent.

Amitav Ghosh has expressed profound concern from time to time that mainstream fiction represents very little of the ecological disasters, which are rather relegated to the peripheral genres of science fiction, gothic or romance, and to modes that are fantasy, surrealism, or magic realism, but scarcely realism:

> When I try to think of writers whose imaginative work has communicated a more specific sense of the accelerating changes in our environment, I find myself at a loss; of literary novelists writing in English only a handful of names come to mind: Margaret Atwood, Kurt Vonnegut Jr, Barbara Kingsolver, Doris Lessing, Cormac McCarthy, Ian McEwan and T Coraghessan Boyle. No doubt many other names could be added to this list, but even if it were to be expanded to 100, or more, it would remain true, I think, that the literary mainstream, even as it has become more engagé on many fronts, remains just as unaware of the crisis on our doorstep as the population at large.[1]

He also says that 'fiction that deals with climate change is not always taken seriously by serious journals' and that in the 'literary imagination, climate change is relegated to the realm of extra-terrestrials or interplanetary travel.'[2]

This point has been reiterated by him in *The Great Derangement: Climate Change and the Unthinkable* (2016):

> That climate change casts a much smaller shadow within the landscape of literary fiction than it does even in the public arena is not hard to establish. To see that this is so we need only glance through the pages of a few highly regarded literary journals and book reviews, for example, the *London Review of Books*, the *New York Review of Books*, the *Los Angeles Review of Books*, the *Literary Journal* and the *New York Times Review of Books*. When the subject of climate change appears in these publications, it is almost always in relation to non-fiction; novels and short stories are very rarely glimpsed within this horizon.
>
> (9)

Notwithstanding the mistake Ghosh makes in stating a very low count of fiction as regards climate change, his concerns are serious enough to serve as some kind of a wake-up call for the literary writers, readers, and critics. They may have no role to play in the big debates mostly raised by environmental scientists, bio-scientists, geneticists, geologists, economists, corporate honchos, diplomats, policymakers, government, and non-government climate watchdogs, etc. who participate in them with much sound and fury. The literature people, however, clearly observe through a confusing mass of scientific data and statistical calculus, hyperbolical statements, rants, terrifying dooms-day prophesies a clear theoretical shift that has taken place from the issues of anthropocentrism and its corollary, i.e., nature–culture division towards Anthropocene. The shift entails a systemic rethinking of the ontology of nature and redefining of human–nature and subject–object relations within a new conceptual framework that problematizes the Descartean ontology, its epistemology of matter–mind dualism, and the resultant distinctiveness of human selfhood.

It was Aristotle who in *Politics* (Book I, Chapter 8) had said that 'nature has made all things for the sake of man.' According to Alder and Wilkinson, 'Medieval Christian doctrine reinforced the same view on the basis that our rationality and immortal soul give us the right to exploit nature and our fall from grace gave us the necessity to do so' (51). But later this view systematically grew as an anthropocentric premise under the influence of the Enlightenment philosophy: that nature, an ontologically self-evident concept and independent of the agency of humans, is still subjected to human will and rational order. Left to itself, nature has no history – so went the belief. For a long time since Giambattista Vico to R.G. Collingwood, the philosophy of history had contained the fundamental idea that human agency and purpose unfolded only through history, beyond which lay nature without much meaning for the humans. At times, as in nature writing, nature was invested with principles and laws of its own that were not affected by history. Nature was considered wild, mysterious, and indifferent to human fate as much in the novels of Thomas Hardy as in Robert Frost's poems. Against this background of nature being separate from history and culture, one can understand the theoretical richness of Anthropocene that is evident from the fact that its historical mapping of the age of the bio-system on a mega geological scale controverts the aprioristic idea underpinning nature.

The stratigraphic periodization of the age of the earth has produced a certain amount of historicity of the processes and forms of life that redefine the debate of nature versus culture insofar as it undercuts the privileged cognitive position of man that Descartes, and in many ways, Hume and Kant, had upheld. Kant helped us to understand how nature and its knowledge were constitutive of human subjectivity. On the one hand, Kant refuted the subjectivist ontology of Descartes that completely ignored objectivism of the phenomenal world, and, on the other hand, he also challenged the empiricism of Hume. His focus on how knowledge of nature is possible led him to postulate nature as a lawful system of order and of causality, governed not by empirical principles, but according to universal laws that exist a priori. Interestingly, even as Kant might have attributed the aprioristic qualities to the universal laws of nature, he at the same time held that the idea of nature's lawfulness and universality were part of a transcendental analytic and 'a priori human knowledge,' which gesture towards the a priori presence of a full-fledged, robustly sentient transcendental subjectivism.[3] In his *Critique of the Power of Judgment*, Kant also attributed the faculty of reason and reflective judgement through which the laws of nature could be known.[4] These philosophical ideas suggesting how nature was epistemologically constructed and controlled by human knowledge affirmed the centrality of human beings and the superiority of human knowledge over nature. The upshot of all this is anthropocentrism.

The sense of certainty, authoritative gaze, and egotism that attend upon anthropocentrism begin to diminish once 'man' is divested of the metronomic authority to define time; and the earth takes over, as it were, to define time and temporality stratigraphically through the rock layers. The Kantian beliefs that time and space are but forms of human intuition and its philosophical legacy suffered a great setback, for the human being is in truth found to be subjected to both space and time that exist independently of subjective consciousness. The post-lapsarian story of the Bible still underscoring the supremacy of humans over plants and animals has been given the lie by geosciences. The knowledge about the origin of the *Homo sapiens* going back to about 315,000 years ago, and therefore considered very recent compared to the earliest forms of life (which is 3.7 billion years old) and the age of the earth (which is about 4.5 billion years old), and above all traceable in the fossil remains and rock strata, reduced the anthropocentric stature of 'man' to just another species among thousands of them, as much perishable as any other. While the biblical and the Enlightenment aura of Man is gone, he is not a 'poor, bare forked animal' to evoke pity, but someone fiendishly arrogant, narcissistic, irresponsible, and self-aggrandizing enough to destroy the same planet that nourishes him and sustains him. As for the planet, it has become terra incognita, once a home from which we have been alienated.

Paul Crutzen's scientific formulation of Anthropocene as a 'human dominated geological epoch' – a clear shift from Holocene – to calculate the enormity of human impact on the earth system is a massive alarm when he says:

> Unless there is a global catastrophe – a meteorite impact, a world war or a pandemic – mankind will remain a major environmental force for many millennia. A daunting task lies ahead for scientists and engineers to guide society towards environmentally sustainable management during the era of the Anthropocene. This will require appropriate human behaviour at all scales, and may well involve internationally accepted, large-scale geo-engineering projects, for instance to 'optimize' climate. At this stage, however, we are still largely treading on terra incognita. (23)[5]

The paradigm shift in the thinking about human habitation within the framework of Anthropocene builds up discursive conditions for four major types of Anthropocene narratives, namely: (1) the naturalist narrative, (2) the post-nature narrative, (3) eco-catastrophist narrative, and (4) eco-Marxist narrative, according to Christophe Bonneuil.[6]

The first category of narrative, the naturalist narrative, a notable one by Dipesh Chakrabarty,[7] states that human pursuit of freedom and progress of the human mind have been a template of civilization since the Enlightenment through technological advancement. This has brought about far-reaching changes in the environmental conditions of the world so much so that the difference between geological history and human history has collapsed, and from being biological species humans have become geological agents 'through their own decisions' (210). He goes on to add, 'The Anthropocene has been an *unintended consequence* of human choices,' and through industrial civilization, we have 'stumbled into it' (italics mine; 217), and that the predicament can be gotten over through collective rationalism by way of global and sustainable environment management. This rationalism is necessary to intellectually conceptualize humans as a species that functions as a geological agent particularly because we cannot phenomenologically experience ourselves as a species in our relationship with the planet. Chakrabarty puts a high premium on our rational capacity to fully comprehend the complex and contradictory linkages between capitalism, the most progressive and modernist of the forces of history, and the unintended environmental disasters. In this narrative human culpability is curiously absent, and it is quite sanguine about the rational capabilities of humans as a collective body to stem the dangers of climate change.

The second narrative that Bruno Latour develops is that Anthropocene has a trajectory towards post-nature. In a lecture delivered in 2011 titled "Waiting for Gaia: Composing the World through Arts and Politics,"[8] Latour argues that the Anthropocene philosophy has replaced the traditional man–nature relationship as has been captured in romantic poetry in terms of the affective power of sublime and vision of eternity. This is because there has already taken place a fundamental disconnect between the two. Humans have become a collective superhuman, considering their enormous capacity for the consumption of resources and energy on the scale of terawatts to sustain the technological civilization and that Anthropocene has been 'anthropomorphism on steroids,' as it were (3). As for the biosphere – the domain of nature – it is no longer framed within a schematized and finite cosmology of the comprehensive and unified universe, but it is a 'scale model that is connected through reliably safe networks to stations where data points are collected and sent back to the modelers.' Nature is not visible from a 'far away point of view' (5), nor is the earth viewed as a global ecological system, but an intricate assemblage of maps, figures, and data through the 'satellites, sensors, mathematical formulae' (7), a corpus of the mediated knowledge of a vast amount of particularities that never lend themselves to any order or scheme, but to an unending array of disparities, contradictions, and transformations. In a schema of actor–network theory, these disparate entities in interaction with the human actants become principles, laws, and theories as much as the humans become actors to invest the principles and laws with programmes and goals that are attributed to nature. This analytic and epistemological perspective of human agency to translate nature and to be translated by it makes us post-natural, as Latour would have it, with the onset of the Anthropocene. In this context, then, nature is not unity; nor indeed can stability or harmony be taken recourse to as riddance from anthropocentric consciousness as the deep ecologists would consider. Nature is, on the contrary, a bewildering network of an incalculably vast multiplicity of entities, agents, intermediaries, interactions, exchanges, and translations. Its modus operandi, although imbricated with human assumptions, intents, goals, choices, and decisions, is still independent of them. Latour, in *Facing Gaia: Eight Lectures on the New Climatic Regime* therefore, invokes James Lovelock's mythological figure of Gaiato to replace nature as a mother, and says, Gaia is not indifferent to humans, but

> quite sensitive to our action and at the same time. She follows goals that do not aim for our well-being in the least. If She is a goddess, She is the one we can easily put out of whack, while She, in turn, may exact the strange sort of revenge.
>
> (9)

Elsewhere, in *Facing Gaia: Eight Lectures on the New Climatic Regime* (2015), Latour presents Anthropocene in a more radical way, redefining it as a post-natural world which is as an obverse of both the universal, law-governed, and unifying system of knowledge that is nature, and anthropologically defined culture as assigning social and theological order, meaning and unity to nature. Differentiating Gaia from nature, he says:

> It is 'nature' that was universal, stratified, incontrovertible, systematic, deanimated, global, and indifferent to our fate. But not Gaia, which is only the name proposed for all the intermingled and unpredictable consequences of the agents, each of which is pursuing its own interest by manipulating its own environment.
>
> (142)

The third, i.e., the eco-catastrophist, narrative serves as a shocking reminder of the nemesis that followed the anthropocentric hubris. Irreversible transformation of close to a half of the earth's surface; unsustainable exploitation of the forest, water, mineral resources, as well as stocks; interference with the course of rivers; anthropogenic erosion of biodiversity were but a prelude to what Clive Hamilton would call 'catastrophism': a large-scale disaster looming large (2017: 26). The trajectory offered is

> towards limits, tipping points, collapse, violence, and wars ... Acknowledging the possibility of a collapse of the industrial way of life and accepting the limits to growth becomes, in the third narrative, an opportunity for a more participatory politics and a new post-growth resilient society where life would be based on a lower and simpler material and energetic base, but with more enjoyable, meaningful and egalitarian communities.
>
> (Bonneuil, 2015, 27)

The fourth, eco-Marxist, narrative, according to Bonneuil, is structured in terms of an asymmetric relation between the expansion of capital and divergence of wealth between nations and social groups. Some scholars use the term 'capitalocene' to refer to the capitalist expansion of the empires and colonization that began in the sixteenth century. It requires little intelligence to understand how the imperial systems of capitalist expansion and profits from overseas were possible primarily because of the nefarious exploitation of the labour, land, and natural resources of the colonial subjects by the colonizers. Bonneuil writes:

> Economic historian Kenneth Pomeranz's path-breaking work (2000) has shown that the control of millions of American 'ghost hectares' – the slave-produced cotton imported by England in 1830 that saved 9.3 million domestic hectares of pasture and hay for production of an equivalent amount of fibre from sheep's wool – played a major role in Britain's economic take-off.
>
> (29)

The story of inequity in economic relation between human groups and nations and exploitation and unequal access to nature are ethically charged issues that are elided by the larger Anthropocene narratives of the unrestrained exploitation of natural resources, carbon energy, destruction of biodiversity, and the resultant catastrophe of climate change. Although Dipesh Chakrabarty does acknowledge this fact, he still argues that the anthropogenic disaster cannot be explained solely in terms of the expansion of capital because:

> While there is no denying that climate change has profoundly to do with the history of capital, a critique that is only a critique of capital is not sufficient for addressing questions relating to human history once the crisis of climate change has been acknowledged and the Anthropocene has begun to loom on the horizon of our present. The geologic now

> of the Anthropocene has become entangled with the now of human history.
>
> (212)

To a student of the Humanities, and particularly literature, what strikes most about the geological scientific discourses and environmental disaster is the deflated image of Man, shorn of grandeur and glory, never like primordial Noah, ordained by God to be the preserver, but a very recent arrival in the pageantry of life forms. He is puny but infinitely fiendish, immeasurably rapacious, and abjectly foolish so as not to know the enormity of ruin he has brought upon the planet. Now the chickens have come home to roost, and what is evoked by Anthropocene could well be 'anthropos-sin'; a terrible sense of guilt for the anthropos (human collectivity) and an ethical burden humans have to carry, and salutarily so, being made to be responsive to as well as responsible for the environment and the fellow creatures. 'After such knowledge, what forgiveness?'

What is also vexing is that Anthropocene is no longer an external and transcendental nature out there under our gaze, to be tamed, studied, or aestheticized, but a bewilderingly complex field of interaction of entities and agents, human actions on the environment, and the consequences thereof. These are not interrelated as causes and effects relation in a linear mode but entangled in never-ending loops with one another. Human action is not governed by free-will and the independent choice of a cogitating self, but only by way of reaction to adapt and survive unanticipated consequences of some previous actions far removed in time and space. We are surrounded by hyper-objects, as Timothy Morton would point out to us, some of which are the products of our actions, such as radiation, hydrocarbons, and global warming.[9] Anthropogenic these are, undoubtedly, but inadvertently so, to take us by surprise. We might even add to the list the latest Covid-19 virus and its recent mutant – popularly called Covid-20 – that are our own unintentional making and already part of the human life system now. These are a reality of our corporeality, a mighty force to change human history, but we have no knowledge of its genesis and provenance. It is not *ab inito*, but *de novo* in our cosmology. Like all other hyper-objects, it is local in its repercussion, and yet pervasive globally as a presence, mixed up inseparably with countless other entities in the biotic phenomena of food chains and life cycles that are basic to the conditions of life on the earth.

The environmental problem so outlined suggests what students of literature have been doing and might do in several ways, which I will very briefly reflect upon in the remaining part of this introductory piece. One of them is how to write the Anthropocene as a literary artist or critic or historians or anthropologist in a non-scientific, humanist register, not just articulating the human predicament of being at the receiving end after millennia of biotic misdeeds, but also discursively engaging non-anthropocentric perspectives and countering the anti-humanist thoughts embedded in them. In their introduction to a special issue of *Minnesota Review* for the year 2014 on the topic 'writing the Anthropocene,' Boes and Marshall say that writing highlights how Anthropocene is 'something actively shaped and created through acts of human inscription' rather than 'simply something that is written about' (64). In fact, writing, as we know, in a gerundive mode, without a subject and preposition, has a distinctive post-structuralist flavour, for écriture – à la Roland Barthes and Jacques Derrida – is an intransitive act, self-directed and a differance, from which a figural immanence of the Anthropocene emerges through a self-reflexive mode. The use of *ecodiegestic* rather than *ecomimetic*[10] modes of the representation of nature, adoption of the narrative technique of prosopopoeia, featuring of cyborg-like subjectivities, and working out of a constant back and forth movement through time and space, refraction of human and historical temporalities from the perspective of the geological counterparts are interesting theoretical possibilities of Anthroposcenic inscription as explored by contemporary fiction. One might cite the example of Jeanette Winterson's *Stone Gods* (2007) in which Spike, a Robosapien (a robot with artificial intelligence) as a non-human subjectivity, negotiates humanism through the positions of alterity such as feminism and lesbianism in technologically altered and post-apocalyptic eco-space of the Blue Planet, within the discourses of capital, technology, and power.

One important question that comes to mind is: Can humanism, which is the essence and spirit of literature, be consigned to the dustbin in the face of the posthumanist crisis of the Anthropocene? I am deliberately using the words essence and spirit that are much maligned in the contemporary posthumanist critical discourse. To my mind, it is not possible to write like an animal or tree, since writing is a human activity and reflection of human consciousness; but one can nevertheless be self-reflexive enough to resist the preponderance of human viewpoint and centredness and develop a new non-anthropocentric cosmology to engage with the non-human forms of life in empathy like David Abram, who situates, among a host of other things, human subjectivity and consciousness within the framework of reciprocity and relatedness in corporeal terms:

> Nourished and sustained by the substance of the breathing earth, we are flesh of its flesh. We are neither pure spirits, nor pure minds, but are sensitive and sentient bodies able to be seen, heard, tasted, and touched by the beings around us.
>
> (63)

In fact, for humans to experience a bat's subjective consciousness is a phenomenological impossibility, says Thomas Nagel, since it is a mammal with its bio-sensory apparatus uniquely designed for echolocation, and that its consciousness cannot have any reductive, materialist explanation. In his article "What is it like to be Bat?" Nagel aptly argues that

> the members of radically different species may both understand the same physical events in objective terms, and this does not require that they understand the phenomenal forms in which those events appear to the senses of members of the other species.
>
> (Nagel ,1974 445)

This being an incontrovertible fact, we post-Enlightenment humans – now sufficiently chastened – can at best assume a cognitive stand that is still profoundly affective concerning the bio-system in the planet. Eco-literary studies can be very edifying for us in this ethical project. We can be posthuman in our analytical approaches and methods within the broad conceptual frame of Anthropocene, and yet we would hold on to our humanist intentions and remain committed to our humanist goals.

Anthropocene fictional works have emerged in a fairly good number in recent years. *Anthropocene Fictions: The Novel in a Time of Climate Change* (2015) by Adam Trexler informs us of this. He claims to have estimated as many as 150 novels about climate change – the major theme of Anthropocene – consisting of works of not just established 'literary' authors, but also middle-brow and critically acclaimed younger authors and those of science fiction who has also written on Anthropocene. Stating that 'there is a [sic] considerable archive of climate change fiction' (7), he charts some of the early works of fiction through the early 1950s and 1960s on greenhouse emissions, terraforming, and long-term shifts in the earth's climate and moves on to survey the thematic changes in fiction throughout the last decades of the twentieth century and the early part of the current century. He mentions that there was

> a spike in publications around 2008, likely due in part to George W. Bush's reelection in late 2004 when there appeared to be little hope of American leadership on environmental issues. Many of the novels featured in this book were written in this period … Since this period, there has been a steady flow of excellent novels about anthropogenic global warming.
>
> (Trexler, 2015, 8–9)

With an astoundingly wide breadth of narratives, the novels raise critical issues for Anthropocene studies such as the historical development of the themes and forms of the climate novel; the limits of the representability of climate change in literature; the significant tropes deployed to articulate different aspects of climate change that could not have been articulated within the conventions of realism, science fiction, or dystopias; the strategies to represent, or even reconfigure, the politics of global warming; the demands climate change as a remarkably broad range of phenomena makes on the novel form; and so on.

As is apparent, Anthropocene extends the frontiers of eco-criticism by bringing to bear upon it the enormity of geological scales on which climate change occurs through the complex enmeshing of human and non-human systems rather than being a consequence of simple human meddling with a pristine nature, which eco-criticism was preoccupied with. As Adam Trexler would say, 'Ecocriticism emerged in the early 1990s, not as a rigorous disciplinary approach, but rather as an interdisciplinary group of researchers interested in literature, culture, and the environment' to critique environmental issues within the space of the academia and politics being much encouraged by feminism, race studies, and postcolonialism that "gained prominence' and attracted 'significant funding and autonomy on North American campuses' (17). Although it would be farfetched, perhaps, for Trexler to dub eco-criticism as some sort of a derivative discourse, there is an undeniable truth about eco-criticism paving the way towards a more systematic and rigorous probing of the anthropogenic process of climate change that the Anthropocene has called into play in the contemporary environmental discourse a whole baggage of new theoretical concepts from geology, culture, climatology, corporate finance, development studies, politics, nuclear and geophysics, biology, genetics, and the like. The epistemological profusion, as well as confusion, the fallacy of pure ontological categories, ethical dilemma, and rhetorical subterfuge, has made the scenario of human agency and responsibility very complex as regards climate change, particularly when a bio-system has been conceived of from where the human beings have been de-centred. But this is certainly at a far end of the trajectory that starts from nature writings and eco-criticism as influenced by the American Transcendentalists and the English Romantics. At the far end, however, the Anthropocene does not supplant eco-criticism. Rather, as a 'threshold concept,' according to Timothy Clark, it not merely gestures towards the limitation of eco-criticism: mostly the naïve assuredness of the green moralism the latter embodies as a communal imaginary against the impersonal scale effects of the behemoth Anthropocenic forces but also suggests what it might do in the future. In concluding his book *Ecocriticism on the Edge: The Anthropocene as a Threshold Concept* (2015), Clark says:

> Ecocritics have been responding to the call for a hypothetical future phase of the Anthropocene, that of an epoch of humanity as the just and responsible steward of the Earth. What else is there to hope or work for? Yet the Anthropocene entails effects that touch on the viability of ecocriticism itself as a possible force of significant change. Environmental readings of literature and culture may need to engage more directly with delusions of self-importance in their practice, keeping alert to the need for more direct kinds of activism. The more complex and even opaque the overall context, then the more any specific framing of it must drift towards simplification. This is a challenge for any sort of activism, but especially for one that limits itself to the realm of cultural representations.
>
> (198)

The chapters in the present volume titled *Interrogating Eco-Literature and Sustainable Development: Theory, Text, and Practice* turn attention to sustainable development, which is one of a whole slew of damage control measures contemplated as the Anthropocene has panned out a climatic disaster of inordinate scales. A secure future for the continuance of the earth system emphasized by the Brundtland Commission Report of 1987 was thought to be achievable through optimal use of resources and by ensuring healthy and lasting interdependence between economy and environment. Since the Anthropocene involves comprehensive knowledge of such factors as water cycles, weather, biosphere, climate system, biochemical cycles, sediment system, and the ways these are structured and processed on the mega scales of geological time and space, it necessitates an integrative approach to balance economic growth with environmental protection through a democratic green politics, of a somewhat Latourian kind, that accommodates the interests, knowledge systems, and environmental practices of the common people as much as the scientific knowledge of the geologists, biologists, climatologists, geneticists, and various other types of environmental experts.

Scholars, not merely from the discipline of literature, but also those of the Social Sciences, have contributed chapters to this volume, approaching different aspects of the anthropogenic crisis in the bio-system and responding to the threat to life on the earth from an interdisciplinary perspective, although with a distinctly humanist accent. Sonam Jalan studies the representation of climate change in the narrative of Cli-Fi that involves a globalist perspective on anthropogenic climate change and a combination of both factual and fictional treatment of socio-economic issues of development, meteorological information, etc. Sarbendu Bikash Dhar makes a statistical analysis of the land use and land cover changes in the adjoining region of the Mangalpur private opencast colliery in West Bengal and the environment-related problems there. He offers a management plan to ensure sustainability. Since the unavoidable fallout of urban development is the depletion of the basic resources of life, Chandan Bandyopadhyay emphasizes the need for the availability of water and maintenance of sanitation as enabling factors of sustainability. As is well known, the debates involving the various corporate, government, and non-government interest groups over the issue of sustainable development and ethics of responsibility towards the bio-system as a whole, and an efficient eco-politics by the states, have been the staple of the Anthropocene discourse. Abhishek Bhattacharya and Sudipa Chowdhury address the fraught relationship between economic development and environmental depredation that leads to the iniquitous divide between the propertied and the pauperized as in Debesh Roy's *Tista Parer Brittanto*, Charles Dickens's *Oliver Twist*, and Aravind Adiga's *The White Tiger*. They invest in literature the ethical role to question the lopsided development stories in human society in absence of sustainability, the factor to balance plenitude and paucity.

The Anthropocenic impetus to the rethinking of nature and the nature–culture binarism is evident in many chapters, inasmuch as the thrust has by and large been to rethink the dialectics of nature and culture which Shruti Das does in hers. The idea of 'urbanature' explored by Sourav Kumar Nag is quite within the Anthropocene paradigm of environmental studies since it also questions the nature–culture opposition. Equally relevant to the Anthropocenic paradigm is Riman Rakshit's problematization of the naively supposed border between human corporeality and agential consciousness on the one hand, and environment on the other as exteriority in the eco-horror and eco-gothic modes of Edgar Allan Poe's fiction.

Displacement is yet another unwelcome consequence of development, and it is indeed one of the biggest challenges humans have faced in the recent past, owing to several reasons. Sajalkumar Bhattacharya discusses human displacement as an upshot of eco-terrorism on the part of the warring states – by those involved in the Vietnam War or the Gulf War that destroyed the bio-systems and burnt the oil-reserves. In order to problematize the issue of the displacement of people due to environmental depredation, he invokes Hannah Arendt's idea of statelessness and eco-terrorism as compelling reasons for postulating a utopian space that is governed by 'the law of humanity.' Ironical as it might seem, at times eco-conservatism or an overzealous environmental protection programme spins off to the dispossession and displacement of people, which happened at Marichjhapi in 1979 in consequence of the launching of the Sunderbans tiger project there. Sujit Malick discusses in his chapter this problem as well as cultural uprootedness with reference to Amitav Ghosh's *The Hungry Tide*. Ghosh dominates the Indian eco-critical literary discourse in a major way with a distinct postcolonial inflection. While a warm humanist, liberal, and truly cosmopolitan anthropological outlook of Ghosh animates his fiction and lends it a corrective perspective on the parochial and skewed mindsets and values that engender animosity and violence across the globe, the self-same outlook also makes impassioned demands for literary imagination when anthropology embraces the issues of climate change.

Nature, which stands at the centre of ecological imagination as an idea, functions as a dominant trope for delineating human potential for the imagination of beauty and spirituality. In this context, William Blake's statement, 'But to the eyes of the man of imagination, nature is imagination itself' (Letter to Dr Trusler, 23 August 1799),[11] comes to mind. Goutam Buddha Sural brings out the latent spiritual valence in the poetry of G.M. Hopkins in post-Industrial Revolution England, calling it a 'spiritual ecology' that served as a panacea for a profound sense of disquiet the poet had experienced.

Quite often the environmental narrative in a literary mode elicits an affective engagement of mind, showing, as Alexa Weik von Mossner does in her book *Affective Ecologies: Empathy, Emotion, and Environmental Narrative* (2017), how we experience the environment in literature and films at the emotional and sensory level. This work offers us the cue to look at how literary artists imagine the environment as a specular figuration of human emotions and habits of mind. Using Lawrence Buell's idea of environmental evocations implicit in the human imagination of space and the experience of it in affective terms, Pradipta Sengupta explores how space in Nathaniel Hawthorne's *The Scarlet Letter* and the short story "Young Goodman Brown" reflects the mental space and human behavioural patterns. Arunima Karmakar emphasizes in her chapter the value of ethical affective response of humans to the environmental crisis.

When nature offers itself as an analogy to the corporeal economy and destiny of woman, an eco-feminist perspective is built up. Anindita Chatterjee reads Alice Walker's *Colour Purple* as an eco-feminist text, while Mohana Das discusses the theoretical aspects of eco-feminism. Nilanjana Chatterjee (Das) argues how Tagore's play *Red Oleanders* (1924) anticipates social eco-feminism by appearing half-a-century prior to Françoise d'Eaubonne's *Le Féminisme ou la Mort.* Saikat Chakraborty's chapter makes an eco-feminist study of Harold Pinter's women characters.

Climatic crisis and the distressful environment issues have either been written about in the discursive mode, notwithstanding the accompanying illustrations at times or shown in a purely visual and non-discursive mode as in films and television shows. But lately, these have been put forth through hybrid texts such as comics and graphic novels, involving inscriptions and images. In his chapter, Pinaki De studies Nick Hayes's refreshingly new take on "The Rime of the Ancient Mariner" that recodes the thematically memorable scenes of Coleridge's original, including the albatross, in terms of horrifying environmental pollution visuals. He goes on to discuss the French cartoonist Philippe Squarzoni's *Climate Changed* as a crash course on science that ranges over a whole lot of issues from how carbon emissions in our earth causes atmospheric warming to the benefits as well as pitfalls of our renewable energy options. Two other graphic texts, namely Orijit Sen's *River of Stories* and Sarbajit Sen's *Carbon Chronicles*, that address the ecological problems in the Indian context also have engaged his critical attention. The Latourian quasi-objectival and hybrid status of the categories of environmental truth, he argues, match well the mixed mode of visual and diegetic signification and communication by the graphic novels/comic texts.

Since the environmental studies and Anthropocene studies constitute an ever-expanding domain of knowledge about assessing, measuring, and remedying the environmental crisis in its myriad aspects, no volume of work at any given point of time, howsoever updated and arrayed with multidisciplinary disciplines, can claim itself adequate and final. Further, as discussed earlier, Anthropocene as a paradigm shift also necessitates rethinking the old issues and articulating them in a new key. The present volume of chapters is well aware of its revisionary purpose and conscious of its limitations.

Notes

1 See Amitav Ghosh's "Where Is the Fiction about Climate Change?" *The Guardian*. Fri 28 October, 2016. https://www.theguardian.com/books/2016/oct/28/amitav-ghosh-where-is-the-fiction-about-climate-change-.

2 Ghosh's interview with Vidya Venkat, titled "The Chains of Causality Which We Cannot See". *The Hindu*. July 19, 2016. https://www.thehindu.com/opinion/interview/%E2%80%98Chains-of-causality-whose-ends-we-cannot-see%E2%80%99/article14497671.ece

3 For a priori human knowledge, see Kant's introduction to *Critique of Pure Reason* (1780) in *The Philosophy of Kant*. Ed. Carl J Friedrich. New York: The Modern Library, 1949. 24–39.

4 See Michel Friedman's *Kant's Construction of Nature: A Reading of the Metaphysical Foundations of Natural Science*. Cambridge University Press, 2013.

5 Crutzen, Paul J. "Geology of Mankind", *Nature*. 415.6867(2002): 23.

6 See Christophe Bonneuil's "The Geological Turn: Narratives of Anthropocene" in *The Anthropocene and Global Environmental Crisis*. Routledge, 2015. 15–31.

7 See Dipesh Chakrabarty's "The Climate of History: Four Theses". *Critical Enquiry*. 35. 2 (Winter 2009): 197–222.

8 In "Composing the Common World through Arts and Politics," a lecture that Bruno Latour delivered at the French Institute in November 2011, he emphasized that the ecological issues are enmeshed with contradictory forces and interests, and hence cannot be managed or traced by any particular institution. See http://www.bruno-latour.fr/sites/default/files/124-GAIA-LONDON-SPEAP_0.pdf

9 See Timothy Morton's *Hyperobjects: Philosophy and Ecology after the End of the World*, 2013. In this book, he develops an object-oriented ontological scheme from which human being has been de-centred. He explores the links between objects and ecology during the crisis-ridden epoch of Anthropocene and problematizes the ecological hermeneutic, especially because we are surrounded by the hyper-objects like Styrofoam, Plutonium, and global warming caused by fossil-fuel consumption, although we are not able to fully experience and comprehend their inherent qualities or provenance in space and time, distributed as these are in space and time on a much larger scale relative to humans.

10 In *Hyperobjects: Philosophy and Ecology after the End of the World*, Timothy Morton talks about ecomimesis as a conventional mimetic, analogical representation of nature, a talking about nature realistically. But ecodiegesis, the term coined by Boes and Marshall (2004) means 'showing' nature and 'giving a voice to the planet' (64). Also see Jennifer Peterson's essay "Ecodiegesis of Nature on Screen" (2019) for an elaborated meaning of ecodiegesis as a scenography of nature, a cinematic artifice of visual nature that is antirealist and therefore anti-anthropomorphic.

11 See *The Letters of William Blake*. Ed. Geoffrey Keynes. Macmillan, 1956. https://archive.org/stream/lettersofwilliam002199mbp/lettersofwilliam002199mbp_djvu.txt

References

Alder, John and David Wilkinson. *Environmental Law and Ethics*. Macmillan, 1999.

Boes, Tobais and Kate Marshall. "Writing the Anthropocene: An Introduction." *Minnesota Review*. 83 (2004): 60–72.

Bonneuil, Christophe and Jean-Baptiste Fressoz. (trans. By David Fernbach). *The Shock of the Anthropocene*. Verso, New York, 2017.

Clark, Timothy. *Ecocriticism on the Edge: The Anthropocene as a Threshold Concept*. Bloomsbury, 2015.

Hamilton, Clive. *Defiant Earth: Fate of Humans in the Anthropocene*. Polity Press, 2017.

Latour, Bruno. *Facing Gaia: Eight Lectures on the New Climatic Regime*. Polity Press, 2015.

Morton, Timothy. *Ecology without Nature: Rethinking Environmental Aesthetics*. Harvard University Press, 2007.

________. *Hyperobject: Philosophy and Ecology after the End of the World*. University of Minnesota Press, 2013.

Peterson, Jennifer. "Ecodiegesis: The Scenography of Nature on Screen." *JCMS: Journal of Cinema and Media Studies*. 58.2 (2019): 162–168.

Thomas, Nagel. "What is it like to be Bat?" *The Philosophical Review*. 83.4 (1974): 435–450.

Trexler, Adam. *Anthropocene Fictions: The Novel in a Time of Climate Change*. University of Virginia Press, 2015.

Weik von Mossner, Alexia. *Affective Ecologies: Empathy, Emotion, and Environmental Narrative*. Ohio State University Press, 2017.

Part I

Through Various Lenses

Theorizing Ecology

1 Tagore's *Red Oleanders*

Tracing an Indian Root of Socialist Ecofeminist Drama

Nilanjana Chatterjee

I

Introduction

Ecofeminism (spiritual ecofeminism in particular) tends to homogenize and simplify the identity of an ecological reformer based on essentialism and biological determinism. However, in the 1990s, socialist ecofeminism attempted to restore the genderless self of human commitment to ecological reformation wherein men aren't stereotyped as inherently repressive and/or oppressive and women as preserver and/or giver. It is a social campaign led by humans, irrespective of gender, against the four strands of modern capitalism – racism, naturism, sexism, and classicism. In brief, socialist ecofeminism is the acknowledgement, both in activism and theory, of the vital relationships between nature, humanity, and society. Although Tagore's play *Red Oleanders* appeared in 1924 and provides an instance of an awareness and a vision which bear close resemblance to the basic tenets of socialist ecofeminism early on, sufficient scholarly analysis is yet to be done in this area. The present chapter, therefore, is a modest attempt to address this gap. While the first section (briefly) traces the socialist ecofeminist origin in Indian classical and medieval drama and situates Tagore's *Red Oleanders* as a significant socialist ecofeminist expression within Indian ecofeminist theatre, the second segment considers the existing critical engagement in the context of ecofeminist readings of *Red Oleanders* (and thereby justifying the necessity of the socialist ecofeminist approach to the text). The third section closely examines the power structures of the Yaksha Town in *Red Oleanders* to study its complex mode of operation (geographically, ideologically, and ontologically) and to identify the Nandini–Bishu–Ranjan triad as genderless propagator of socialist ecofeminism, committed to strike back and restore balance between nature, humanity, and society.

DOI: 10.4324/9781003383970-3

I

In a talk in 1924[1] (published in *The Visva-Bharati Quarterly*), Tagore argued that *Red Oleanders*' theme – involving unscrupulous capitalism, environmental exploitation, and the importance of human relationships – was not an obscure one, as the Occidental readers accused him of. Some 90 years after the publication of *Red Oleanders*, it seems past time to review the text's contribution to socialist ecofeminist thoughts to trace one of the Indian roots of a growing world movement. Tagore himself, however, believed that the play's theme involving the unscrupulous pursuit of wealth at the expense of human relationships and environmental disaster was indeed apprehended by as ancient a poet as Valmiki in *The Ramayana*. Indian ecofeminist consciousness – genderless (combining the aspects of the *purusha* and the *prakriti*[2] within the dynamics of nature), nonprescriptive, rational (determined by the Dharma and Karma relationship based on the law of cause and effect), strategic and righteous (not religious) – probably was rooted in the Vedic culture (Krishna Nanditha, 2020) wherein Indra, Agni, Vishnu, Pushan, Ushas, Dyauspitr, Prithvi, and rivers[3] were considered as integral to sacred environmental knowledge which intersected with environmental history (environmental history being inclusive of various natural calamities such as earthquake, drought, or flood). Indian literature, since then, is an anthropocentric representation of these human engagements with environment and nature (for example, 'Nadistuti Sukta,' 'Prithvi Sukta,' and Shanthi mantra in the *Rig Veda*, the *Atharva Veda*, the *Yajur Veda*, respectively) wherein preservation/destruction of air, water, and land lead to a consequential salvation/damnation. Krishna in the *Bhagvad Gita* represents earth, water, and air as projection of his sacred self. Sanatana (meaning eternal) Dharma (meaning righteousness), according to Hindu, Buddhist, and Jain culture, has explicated the interconnectedness of duty, cosmic law, and justice which together ensure sustainable welfare of human, animal, and plant life on the earth and balancing of the Pancha Maha-bhoota.[4] The *Purana* acknowledges nature as integral to inter-being – all species are equal in the eyes of the Creator wherein human control is discouraged. *Aindu Tinai*[5] represents geographical landscapes as realms divided between Lord Murugan, Lord Krishna, Lord Indra, Lord Varuna, and Goddess Kotravai. Traditional Indian plays – both classical and medieval – reflect a similar genderless and rational ecofeminist righteousness early on.

Catherine Diamond (2017) in "Four Women in the Woods: An Ecofeminist Look at the Forest as Home" identifies Sakuntala in Kalidasa's fifth-century Sanskrit play *Sakuntala* (besides Titania in *A Midsummer Night's Dream*, Neang Seda in the *Reamker*, and Rosalind in *As You Like It*) as one of the four female characters who experiences the forest space without looking at it as a symbolic projection of feminine body within a patriarchal society. Kalidasa extracts an episode from the *Mahabharata* to explore the possibilities of meaningful human–nature interconnectedness (and its absence).

Diamond (2017) points out that the orphaned child – rescued by birds (*sakuna*) and raised by a hermit in a forest – extends familial bonding with forest animals and plants, representing the lived life of a forest denizen who shares an ambivalent relationship with her space and place: Adult Sakuntala feels discontent with the forest environment but also protects her putative society: 'Withdraw your [King Dusyanta] well-aimed arrow! Your weapon should rescue victims not destroy the innocent' (Diamond, 26). When Sakuntala takes leave of the hitherto lived forest (as opposed to conceived or perceived forest space) to go to the capital, her family members respond: 'Grazing deer/ drop grass, / peacocks/ stop dancing, / vines loose/ pale leaves/ falling/ like tears' (Diamond, 26). King Dusyanta's coldness and inhumane treatment towards Sakuntala represents universal human cruelty which is further projected in his troubled relationship with nature and agricultural conditions. However, later in the play the King reflects Sakuntala's philosophy of preserving the forest and of looking at it as an experiential space. He reprimands his biological son (without knowing his real identity): 'why do you violate the sanctuary laws/ and ruin the animal's peaceful life, / like a young black snake in a sandal tree?' (Diamond, 27) wherein his transformed philosophies in the context of forest space are revealed. The classical play represents early on elements of socialist ecofeminist possibilities as the text problematizes lived human–nature connections by exploring fluid and complex relationships between humanity and environment within a given society rather than narrowing down the lived space into a pair of binaries – forest/ court; Sakuntala/Dusyanta – assigning fixed roles to each. Indian medieval plays – consisting of traditional folk theatres from various parts of India ('Ramleela' and 'Raasleela' in North India, 'Veedhi Natakam' in Andhra Pradesh, 'Yakshagana' in Karnataka, 'Bhavai' in Gujarat, 'Nautanki' in Uttar Pradesh, 'Yatra' in Bengal, 'Ankiya Naat' in Assam, 'Krishnattam' in Kerala, etc.) – like the classical plays heavily depended on the *Ramayana*, the *Mahabharata*, and the Puranas for their thematic contents. These theatrical forms were rooted in indigenous locality, ecology, and myth, upholding traditional doctrines of human–nature–society interconnectedness and/ or disconnectedness, as represented in the Indian epics and the Puranas. The projection of experiential and mythical forests (*tapovana*, *mahavana*, *shreevana*); the symbolical and material allusions to botanical, zoological, and medicinal aspects of the forests; and the enactment of complex human association/dissociation with these aspects of nature within broader ecological context played as a metronome throughout the traditional folk plays. While ecofeminism engages in identification and restoration of women and environmental oppression in the hands of patriarchal cultures, ecofeminism with social consciousness emphasizes on visibilizing the latent interconnections between all forms of oppressions that operate within a capitalist society. Armbruster (1996) observes, 'central to the ecofeminist agenda is the goal of individual, social, and ideological change – specifically, change that will improve the cultural standing of women and nature' (97). Nature/

female and culture/male dualisms not only perpetuate the trope of femininity and masculinity but also fail to sufficiently problematize the dissemination and access to environmental knowledge.

II

Murphy (1998) in "The Women Are Speaking" asserts, 'with literature, readers need to be asking themselves constantly what the texts in their hands can offer to enhance the theories that shape their lives' (24). The present chapter, with a similar political commitment, intends to explicate the anti-essentialist strategy of socialist ecofeminism (as opposed to spiritual ecofeminism), which runs all through the text and disrupts all forms of binary division to mobilize social changes. Social ecologist and feminist Biehl (1991) in *Rethinking Eco-Feminist Politics* charges ecofeminism for its unrealistic approach to an arbitrary bond between women and nature (for example, Sherry Ortner's ecofeminist perspective in *Is Female to Male as Nature Is to Culture?*). Cecile Jackson (1993), Meera Nanda (1991), and Bina Agarwal (1992) too have criticized this ecofeminist approach for it fails to consider the material sphere of exploitation and domination. Agarwal (1992) points out that the approach neglects the 'interrelated material sources of dominance based on economic advantage and political power' (122). Moreover, Greta Gaard (2011) in "Ecofeminism Revisited: Rejecting Essentialism and Re-placing Species in a Material Feminist Environmentalism" criticizes ecofeminism for its gender discriminating (propagating dualism based on gender differences) and gender stereotyping attributes (ascribing women a given set of roles). Biehl objects to the stereotyping of women as nurturer and earth as female as regressive in nature.

The present study therefore does not attempt to stereotype the roles of femininity/masculinity by labelling Nandini in *Red Oleanders* as some female mystic nurturer and the King as some patriarchal capitalist based on biological determinism. This is exactly what socialist ecofeminism intends to undo within the dynamics of a broader spectrum of academic discourse by emphasizing the fluidity and heterogeneity of human relationship with nature and society. The present segment visits the existing scholarly engagement in relation to ecofeminist reading of Tagore's *Red Oleanders* to contextualize the need of the present study. Asha Kuthari Chaudhuri's "Old Text, New Theory: Reading Tagore's *Red Oleanders* through Ecofeminist Lenses" (2015) is probably the most significant critical analyses of *Red Oleanders* in the light of ecofeminism – rich in theoretical perspectives and thematically interlocked structure. Chaudhuri (2015) uses ecofeminism as a tool to read the text wherein she identifies Nandini as 'naïve female with a jouissance that is at once able to override the stupor of both the rulers and the ruled with her vitality and unbridled joy' (61). She acknowledges early on the ecofeminist dependence on 'logic of domination,' which advocates the 'master model' that associates women with nature, the material, the emotional, and the particular, and men with culture, the nonmaterial, the rational, and the abstract. In doing so, Chaudhuri (2015) perhaps makes an essentialist reading of the Nandini–Bishu–Ranjan feminist project (Adichie's *We Should All Be Feminists* interprets feminism as an inclusive awareness process) of subversive resistance against state machineries in Yaksha Town by labelling the trio's rebellious act as essentially 'feminine principle.' The essay is a (fe)male discourse as Ranjan has been labelled as 'the magical lover of Nandini' representing the lost possibility of genderless ecofeminism, albeit being painfully aware of the pitfalls of essentialism. The essay even refers to Davion and other ecofeminists who look at radical ecofeminism as a dangerous feminine trope and trap but perhaps falls into a similar trope/trap of essentialism so much so that the essay neglects sufficient exploration of Bishu and Ranjan's feminist acts of subversions. Moreover, Chandra's unapologetic anti-ecofeminist stance and King's transformation by the end of the play have not been sufficiently problematized. The next segment therefore is a modest attempt to visibilize the heterogeneity and fluidity of ecofeminist project in *Red Oleanders* wherein agents of state machineries are anti-ecofeminist (which too is a heterogeneous and fluid identity).

III

The present segment engages in textual analysis of the socialist ecofeminist elements in the text by underscoring the genderless (and therefore complex) identity of ecofeminist/anti-ecofeminist agencies in a capitalist society at Yaksha Town. Socialist ecofeminism is a recent iteration of social versions of ecofeminism, which according to Ehrenreich is a hybrid of Marxism and feminism and has 'the potential to be greater than the sum of its two parts' (Shapiro and Shapiro, 1979, 207). It is a fact that socialist ecofeminism involves activism as well as ideology. The essay entitled "Ecofeminism: Toward Global Justice and Planetary Health" (1993) by Greta Gaard and Lori Gruen – inspired by the insights of ecology, feminism, and socialism – draws the four sides of the theoretical frame of ecofeminism to problematize the current global situation by meaningfully identifying the intricate dynamics of the present Orwellian environmental crisis and initiating political commitment and social change. The four sides, as drawn by Gaard and Gruen (1993), appear to be interconnected: Firstly, the mechanistic-materialistic model of the universe tends to reduce elements of nature and its relationship with humanity into mere commodification of resources in a capitalist society. Secondly, in such a society, humanity suffers, leading to self and other binaries and thereby creating pockets of inherent power domination. Thirdly, in this repressive social structure, discrimination based on gender becomes one of the most prominent forms of exploitation. Fourthly, capitalism – state-run or otherwise – in the name of growth and development – annihilates animals, plant life, and people for the sole purpose of creating wealth. Ecofeminism, in its ideal form, thus, identifies interrelationships between all forms of oppression. It is a practical movement for social change. In brief, ecofeminism is the acknowledgement, both in activism and theory, of the vital relationships between nature, femininity, and humanity. *Red Oleanders* assembles some of the most insightful advocates of the socialist ecofeminist perspectives and the chapter makes a close reading of these enactments. In doing so, the chapter also seeks to analyze if such a reading can illuminate socialist ecofeminism as a valuable component of literary criticism by acknowledging the fluidity and heterogeneity of the relationship between nature and humanity.

In the play, the King of Yaksha Town, who always operates from behind a screen, exploits nature as well as all possible human resources – of mind, of science, of religion – in the most ruthlessly mechanical way to establish a highly centralized system of oppression and exploitation and add to his ever-growing wealth of gold. Topographically, ontologically, and ideologically, the place is guarded by the King's guards. Into this inert town enters Nandini, an emblem of joy and life. Nandini, her beloved Ranjan, her 'mad one' (24) Bishu, and her loyal Kishore, with their undying love for life, nature, humanity, are naturally misfits in the Yaksha Town. They get locked with the authorities in an unequal fight. But it is (chiefly) through Nandini that the socialist ecofeminist revolution courses through the play, and ultimately wins, though through great grief and sacrifice. To Professor, Nandini is a treasure, not of the dust, but of 'the light which never owns any bond' (4). Antiquarian identifies her as 'a girl wearing a grass-green robe' (31). However, Nandini has been envisioned by her author as a vivacious person whose mind and body are in tune with the specificity of place, i.e., her village which is in harmony with the natural environment. Agarwal (1992) points out that nature, culture, and gender are 'historically and socially constructed and vary across and within cultures and time periods' (123). Tagore's Nandini is also represented as a human determined by place, class, and ethnicity rather than as a homogeneous mystic entity. Even while in the town, she recalls, how boys from her neighbouring villages come to race their boats on the river, in rainy June. Her heart aches for fellow villagers who later become the 'King's leavings' (33). When the autumnal sun spreads the glow of the ripening corn in the air, her heart dances with joy. So, it is quite natural for her – unlike Chandra – to persistently resist the anti-nature deadly socio-ecological situation of the Yaksha Town.

The red oleanders of the only red oleander tree in the horrid Yaksha Town – hidden away behind a rubbish heap – emerges as a significant environmental agenda. Kishore – representing youth literally and metaphorically – spots the tree and brings the flowers to Nandini at great risk. Nandini wears this red oleander on her neck, on her breast, on her arms as a remembrance of Ranjan's love. These few delicate petals of red oleanders guard her from the omnipotent King and his omnipresent surveillance. King finds these flowers to be equally alluring and abhorring: at times, these flowers appear to him as evil stars and he wants to tear them to pieces and at times he desires to have these flowers sprinkled on his head by Nandini so that he could die in peace. There are times when King feels threatened by the red petals as if they are 'evening storm clouds gathering for a night of terror' (34). The red oleanders intensify the inner conflict in King as the destroyer-and-preserver and makes him painfully aware of his anti-eco-ethical self. The flowers alarm him of the limitless possibilities of the latent revolutionary spirit in society and nature. Kundu (2012) in her "Introduction" to *Red Oleanders* observes that the bunch of red oleander flowers can be considered the floral symbol of what Nandini manifests as a human figure. Here, she aptly refers to Pramathanath Bishi's interpretation of Nandini as the essence of life's vibrancy and the flower as the essence of Nandini. Nandini dares to challenge the King but most importantly, she can awaken his dormant conscience and inculcate in him ecosocialist ways of living albeit with great personal sacrifice. King identifies the embedded (but powerful) resistance in Nandini early on: 'Within you [Nandini] there is the hindrance, so strong because so soft' (8). Red oleanders, like Nandini, evoke the indomitable spirit of life above all forces of death. Naturally, when Nandini arrives at Yaksha Town, she gets disheartened by the severe environmental degradation on the landscape with only one red oleander tree.

According to the Professor, the town is a city under eclipse – 'The shadow Demon, who lives in the gold caves, has eaten into it' (4). Here, drunken male diggers – in their repressed rage – tear the earth's surface to pieces to accumulate heaps of gold metal. In the process, Chandra claims, the men have been eviscerated and reduced to machines. But the meaningfully mad one, Bishu, points out that women too have lost their ecoethical consciousness as their 'flowers have faded, and you [Chandra] are all slavering for gold' (14). King, identifying this human-and-nature degeneration, repents at one point as he confesses to Nandini that the flaming thirst of the desert of Yaksha Town licks up one fertile field after another, only to enlarge itself for the landscape has lost its fertility to heal even a grass on it. Nandini attempts to save King from further degradation (both within and without): As she hears the autumnal song coming from the distant fields, she invites the King to participate in the joy of food grain productivity and food security. But to her disappointment, the King fails to respond as his existence is now of a stagnant lake which 'cannot run out dancing, like a frolicsome waterfall' (8). Nandini interprets the socialist ecofeminist qualities of a tiller as she argues with the King: the work of the tillers is much more meaningful, productive, and delightful than the work of the diggers in Yaksha Town. To her, earth has a living heart that yields to the tillers in love and life and beauty, but when the diggers rend the earth's bosom, they invite enormous ills causing immeasurable destruction and degeneration.

In such a mechanistic-materialistic system, wherein production of gold is prioritized over food security, safety, and sustainability, humanity suffers, leading to man's alienation and destruction. Schueler, Kuemmerle, and

Schroder (2011) have identified surface gold mining as a source of deforestation, substantial loss of farmland, and degradation of livelihood:

> Surface mining ... removes vegetation and soils, interrupts ecosystem service flows, and results in inevitable and often permanent farmland loss ... in the developing world often erodes livelihood foundations ... As a consequence, conflicts between communities and mining operators over land use rights are common in many regions worldwide ... and can become a serious threat to development and security.
>
> (528)

A similar risk works in the environment, livelihood, and economy of the Yaksha Town. The town's socioeconomic power game operates through a hierarchical chain with King at the top and workers at the bottom. In the middle, come people like the Governor, the Deputy Governor, the Headman, and the Gosain who implement diverse state machinery to retain their power over the diggers. To Nandini's dismay, within this capitalist patriarchal power structure, the whole city thrusts its head underground, groping with both hands in the dark to dig out dead wealth that the earth has kept buried for ages. Professor explains to Nandini that the purpose of enslaving that dead wealth is to enslave the whole world. Such a totalitarian system of 'grabbing and killing' (8) inevitably generates all pervasive mistrust and meanness. The denizens are either angry, or suspicious, or afraid living in constant fear of being overheard by their superiors who are devoid of the milk of humanity. They are insentient towards change and beauty. From Gokul to King – all refuse to trust what they don't understand. Gokul and Chandra call Nandini a witch and Gokul even wants to burn her alive. The King wants to pluck her out of her beauty, to grasp her within his closed fist, to handle her, to scrutinize her – or else to break her into pieces.

Merchant (2005) observes that socialist ecofeminism considers nature and human nature to be 'socially constructed, rooted in an analysis of race, class and gender' (194), thereby emphasizing the dialectical relationship between 'production and reproduction, and between production and ecology' (195–197). *Red Oleanders* represents the diggers as workers who strive involuntarily all day in the dark underground, and in the evening steep themselves in the denser dark of drunkenness. They exist mechanically in figures and follow one another in rows. The work process is endless, without leisure, somewhat like the calendar that never records the last day. On holidays the workers, like the caged birds, spend their leisure knocking against the bars. However, in an ideal working condition, as Bishu explains, wherein human works in harmony with the environment, the holidays are spent in 'the green of the woods, the gold of the sunshine' (13). But when the work is to eviscerate the environment, 'nature's own ration of spirits' (13) ceases to be. It is then that the inner darkness in the workers craves the wine of the marketplace to lull their ecoethical senses.

The most alarming damage that the power system does to the diggers of Yaksha Town is to ideologically capture them to such an extent that the diggers' willpower to break away from the autocratic rule of the Town is permanently crushed. Bishu explains to Phagulal and Chandra, 'If you go there to-day you will fly back here to-morrow, like a caged bird to its cage, hankering for its drugged food' (14). The High Preacher, Kenaram Gosain, is appointed by the ruling class to brainwash the workers in the name of religion to make them meek and servile. Following the Governor's order, the Preacher instructs Chandra and Phagulal:

> Just think of it, friend 47 V, yours is the duty of supplying food to this mouth which chants the holy name. With the sweat of your brow have you woven this wrap printed with the holy name, which exalts this devoted body? Surely that is no mean privilege. May you remain forever undisturbed, is my benediction, for then the grace of God will abide with you likewise.
>
> (17)

The Preacher's solution to oppression is to repeat aloud the holy name of Hari. To make them disciplined, he propagandizes for a deceiving form of emancipation. He preaches them to follow the path of modest suppression and holy subjugation to the ruling class to restore the class inequality in terms of wealth and power. When Gosain fails to interpellate them, the Governor enters the scene with his blunt ruthless repressive methods. The victimized Wrestler laments to Nandini: 'these demons know the magic art of sucking away not only strength but hope' (35). Nandini challenges Gosain, 'what Good God has so heavily charged you to do to these people' (37). She charges him of deluding hapless people of the town with words, after winning their faith. Not only Religion, but Science and History fail to liberate the denizens of the town as the scholars here mechanically examine nature without environmental commitment. While Professor and Antiquarian represent the unproductive theorization of Science and History, it is Nandini who ushers in a new current in the stagnant darkness. Until Nandini appears on the scene, citizens of the Yaksha Town remain nearly oblivious to their oppression in the hands of power. She influences them, one by one, to long for a change. That she can do so makes her an adequate voice of what we call socialist ecofeminism today.

Ecofeminism is a practical movement for social change and Nandini, the protagonist of Tagore's *Red Oleanders*, seems to anticipate in a way the revolutionary spirit of the movement. She seems to echo the foundational tenet of socialist ecofeminism: Revolutionary social change is required for restoring humanity and ecological balance. As the blooming red oleanders, ecoethical commitment too is the product of human–nature relationship. Nandini's undaunted resistance against all forms of oppression is essential for restoring nature and humanity. Nandini challenges the Governor:

> Because I am a woman, you are not afraid of me? God sends His thunderbolt through His messenger, the lightning spark – that bolt I have borne here with me; it will shatter the golden spire of your mastery. (37)

However, it is interesting to note that unlike much of the anti-essentialist theorizing and activism in the history of ecofeminism, which took place in the early days of the movement, Tagore's *Red Oleanders* does not project gender stereotype of women as being inherently close to nature as opposed to men. Nandini, along with Ranjan and Bishu, represents the eco-sensitive spirit (King, Governor, Gosain, Professor, Phagulal, and Chandra represent the anti-ecofeminist perspectives). Critics like Susan Prentice (1988) argue that humans destroy the earth regardless of gender implying that human can nurture nature regardless of gender. Tagore's Nandini doesn't believe in polarizing the earth into categories. She strongly believes that Ranjan can put a beating heart behind the dead ribs, can bring God's own laughter in the Town, thereby startling the denizens of the lifeless place. When Ranjan is put to work in the tunnels of Vajragarh, he fearlessly introduces among the diggers 'Digger's dance' (29) that makes the digging interesting and stimulating. Bishu, on the other hand, is Nandini's partner in her clinging love for the earth and the sky. In Yaksha Town, a bit of sky survives only between Nandini and Bishu. After Ranjan's death, Bishu and Nandini revolt against the power structure along with the other workers of the Town. Both Ranjan and King are deceived by King's own men. Nandini places the blue bird's feather in Ranjan's crest and declares that Ranjan's victory has begun from that day, and she is its bearer. She prepares herself for the last fight – the fight between King and herself. The ultimate victory is complete when King – transformed in the end – himself comes out, defying the net and his militia, which are now bent on thwarting him. Ultimately the forces of life triumph over the forces of death – a process dreamt by Ranjan and initiated by Nandini, and obstructed by the system, finally emerges triumphant.

IV

In his play *Red Oleanders*, Tagore provides us with an early instance of awareness and a vision that bears a close resemblance to the tenets of socialist ecofeminism. Tagore realized and pointed out in several of his essays and speeches the embedded lacunae of unrestrained industrialism, motivated by the greed for power and wealth, which, he felt, would eventually lead to the 'crisis of civilization' (the title of his last public speech, April 1941). Tagore in "Can Science Be Humanized?" (1933) drew a balance between machine and humanity. There he observed, 'There is no meaning in such words as spiritualizing the machine; we can spiritualize our being which makes use of the machine' (Kundu, xxiv). In a way, *Red Oleanders* is about a sustainable negotiation between nature, humanity, and machine. Moreover, the text emphasizes the genderlessness and the placelessness of Tagore's socialist ecofeminist consciousness in the context of industrial colonization by positing Nandini–Bishu–Ranjan's music, dance, and flowers as the most meaningful language of human–nature coexistence. However, it is absurd to assume that a creative artist and an original genius like Tagore would write a play to illustrate any particular theory. Rather his text is made to embody his vision of life, which is deeply felt and sensitively theorized. It is only subsequently, in the after years, that critics strive to bring new light to bear upon it. The test of a loaded literary text is that it reworks and rejuvenates itself to multiple interpretations across ages.

Notes

1 Talk of Tagore, transcribed by L.K. Elmhirst in 1924, during the poet's visit and stay in Argentina, November–December 1924; published in *The Visva-Bharati Quarterly*, November 1951–January 1952, under the title "Red Oleanders: An Interpretation."
2 While Purusha represents the masculine aspects, Prakriti represents the feminine aspects.
3 Indra is equated with Rain, Agni with Fire, Vishnu with Sun, Pushan with Agriculture, Ushas with Dawn, Dyauspitr with Father of the Shining Ones, and Prithvi with Mother Earth.
4 Pancha Maha-bhoota is a group of five basic elements, which, according to Hinduism, is the basis of all cosmic creations. The elements are Prithvi (the earth), Apas (the Water), Agni (the Fire), Vayu (the Air), and Akasha (the Space).
5 Aindu Tinai is the ancient Tamil literary theory on environmental knowledge.

Works Cited

Agarwal, B. "The Gender and Environment Debate: Lessons from India." *Feminist Studies*. 18.1 (1992): 119–158.

Armbruster, Karla. "Blurring Boundaries in Ursula Le Guin's "Buffalo Gals, Won't You Come Out Tonight": A Poststructuralist Approach to Ecofeminist Criticism." *Interdisciplinary Studies in Literature and Environment*. 3.1 (1996): 17–46.

Biehl, Janet. *Rethinking Eco-Feminist Politics*. Boston: South End Press, 1991.

Chaudhuri, Asha Kuthari. "Old Text, New Theory: Reading Tagore's *Red Oleanders* through Ecofeminist Lenses." *The Politics and Reception of Rabindranath Tagore's Drama: The Bard on the Stage*. Edited by Arnab Bhattacharya and Mala Renganathan, London: Routledge, 2015. 61–68.

Diamond, Catherine. "Four Women in the Woods: An Ecofeminist Look at the Forest as Home." *Comparative Drama*. 51.1 (2017): 71–100.

Gaard, Greta. "Ecofeminism Revisited: Rejecting Essentialism and Re-Placing Species in a Material Feminist Environmentalism." *Feminist Formations*. 23.2 (2011): 26–53.

Gaard, Greta and Lori Gruen. "Ecofeminism: Toward Global Justice and Planetary Health." *Society and Nature*. 2.1 (1993): 1–35.

Jackson, C. "Women/Nature or Gender/History? A Critique of Ecofeminist Development." *The Journal of Peasant Studies*, 20.3 (1993): 384–419.

Krishna, Nanditha. "Sacred Ecology of India (1)." https://youtu.be/t4lnsmAr0oc, 12. 02. 2020. Accessed on 07.05.2021.

Kundu, Rama. "Introduction." *Red Oleanders: A Drama in One Act*. Edited by Rabindranath Tagore, Peacock Books, 2012. vii–lxxxvi.

Merchant, Carolyn. *Radical Ecology: The Search for a Livable World*. Routledge: New York, 2005.

Murphy, Patrick D. "The Women Are Speaking: Contemporary Literature as Theoretical Critique." *Ecofeminist Literary Criticism: Theory, Interpretation, Pedagogy*. Edited by Greta Gaard and Patrick D. Murphy, University of Illinois Press, 1998. 23–48.

Nanda, M. "Is Modern Science a Western Patriarchal Myth? A Critique of the Populist Orthodoxy." South Asian Bulletin. Vol 11, 1991. 110–116.

Ortner, Sherry B. "Is Female to Male as Nature Is to Culture?" *Feminist Studies*. Vol 1 (2), 1972. 5–31.

Prentice, S. "Taking Sides: What's Wrong with Ecofeminism?" *Women and Environments*. Vol 10 (3), 1988. 9–10.

Schueler, Vivian, Tobias Kuemmerle, and Hilmar Schroder. "Impact of Surface Gold Mining on Land Use System in Western Ghana." *Ambio: A Journal of the Human Environment* (201) 40: 528–539. 18 March 2011.

Shapiro, Evelyn, and Barry Shapiro. *The Women Say, the Men Say: Women's Liberation and Men's Consciousness: Issues in Politics, Work, Family, Sexuality, and Power*. NY: Delacorte Press, 1979.

Tagore, Rabindranath. "Can Science Be Humanized?" *The English Writings of Rabindranath Tagore*. Ed. Sisir Kumar Das. New Delhi: Sahitya Akademi, 1996. 665–666.

Tagore, Rabindranath. *Crisis in Civilisation and other Essays*. Delhi: Rupa & Company, 2003.

Tagore, Rabindranath. *Red Oleanders: A Drama in One Act*. New Delhi: Peacock Books, 2012.

2 Dialectics of Nature and Culture

Shruti Das

Nature and culture since the beginning of civilization are in a binary relationship. Nature, or as we understand the non-human world, offers deep ecological wisdom often othered and overlooked by technologically advanced human beings. Ecological researchers bring forth ideas of interconnectivity between the human world and that of non-human nature and seek to establish balance and harmony between the two. The first-generation ecocritics valorized nature writing and generally presumed it to be rooted in transcendentalist thoughts as in the writings of Thoreau, Emerson, and Fuller and in the imaginative language of poetry of the Romantics who either escape into nature or from it. They drew attention to the cultural value of nature writing. We read in Wordsworth's "Tintern Abbey":

> How oft in spirit, have I turned to thee, / O sylvan Wye! Thou wanderer thro' woods/ How often has my spirit turned to thee.
>
> (*Poetry Foundation*, no page)

Consecrating nature as a saviour or nurturer has been common to all nature writers, and ecocritics have been concerned with the cultural and literary value of such writing. Dana Phillips believes that ecocritics, to begin with, had immense faith in all things green and were reactionary. In her book *The Truth of Ecology: Nature, Culture and Literature in America.* she calls this the ecocritic's epiphany and that it

> can be summed up by the propositions (1) that nature, which is refreshingly simple, is good; and (2) that culture, which is tiresomely convoluted, is bad; or (3) at least not so good as nature. And insofar as the ecocritic's epiphany inspires such thoughts, its implications are largely reactionary
>
> (Philips 2003, 3).

DOI: 10.4324/9781003383970-4

Further, the environmental historian William Cronon writes in "Introduction: In Search of Nature." *Uncommon Ground: Toward Reinventing Nature*:

> Popular concern about the environment often implicitly appeals to a kind of naïve realism for its intellectual foundation, more or less assuming that we can pretty easily recognize nature when we see it and thereby make uncomplicated choices between natural things, which are good and unnatural things, which are bad.
>
> (Cronon, 1995, 425–426)

If the history of ecology and human history on earth teaches us anything, it teaches us that nature isn't so easily recognized. In this chapter, I shall pit nature against culture in both its historical and literary context which indeed is a contentious struggle fraught with paradox. I shall historicize and theorize the nature–culture binary while limiting myself to the Indian context so far as contextual/textual analysis is concerned.

Humans have always engaged with the physical environment and nature figures unabashedly as human habitat is limited within human cognition and expressed through human language. Unfortunately, the ecological crisis today mandates the expression of the troubling material expression of modern culture's philosophical assumptions, epistemological convictions, aesthetic principles, and ethical imperatives. Therefore, ecocriticism has evolved to become a theoretical and methodological tool that focuses on the imaginary boundaries between nature and culture. It theoretically and conceptually investigates nature, not as a concept that reinforces but one that challenges established cultural, political, and ethical normativities.

Elements of nature, such as climate, topography, animal and insect life, vegetation, and soils, directly or indirectly shape human activity and productivity. In affecting land-use and subsistence, they help to promote or prohibit specific forms of social structure, economic organization, and belief systems which we call traditional or cultural beliefs. They also foreground non-human characters which are considered insignificant and normally ignored. Alternately, man alters the landscape; fells trees; erodes soils; restricts streams, dams, rivers; and kills off plants that seem troublesome and predatory animals, installing favoured species in their stead. Arnold and Guha hold that the 'awareness of man's dependence upon nature has a long ancestry; but a sense of man as the maker and unmaker of nature has only more recently dawned upon us' (3) and has become an eco-historical concern. '[W]ithin the broad bounds of environmental history, there is a history of the environment as cultural space and ideological artifact, as expressed through the invocation and representation of nature in art and religion, in myth, in ethics and the law'. They observe that it 'has recently become fashionable in the West to explore environmental themes in the "ideology and iconography ... [in] the novels of James Fenimore Cooper, or the poems of

Walt Whitman' (3). While they ignore or undermine the fact that 'the history of the environment, in both its material and perceptual sense, is also a history of popular perceptions and experience, of folk traditions and religious beliefs ...' (4).

Richard Kerridge furthers Cheryll Glotfelty and Lawrence Buell's definition of ecocriticism and defines it as a project that 'seeks to evaluate texts and ideas in terms of their coherence and usefulness as responses to environmental crisis' (Kerridge and Sammells 1998: 5). By implication, this definition challenges dominant modernist assumptions about literature and art as aesthetically and ethically autonomous entities. And such a position entails a critical reassessment of the functional relationship between cultural 'texts' and their material referents, i.e., a re-evaluation of mimesis and representation as core categories of literary and cultural criticism. Another prominent ecocritic Laurence Coupe says that ecocriticism's ultimate objective should be to encourage resistance rather than conservation – 'resistance to planetary pollution and degradation' (Coupe 2000: 4), a politically charged demand that presupposes changes in established patterns of thought and behaviour brought about by an (eco)critical inspection of culture's discursive relationship to nature. As Coupe also insists, 'green studies does not challenge the notion that human beings make sense of the world through language, but rather the self-serving inference that nature is nothing more than a linguistic construct' (3). Armbruster and Wallace believe that the inclusion of urban, ethnic, and national perspectives in an ecological approach to literary and cultural studies is necessary to avoid ecocriticism's theoretical and conceptual self-marginalization in the larger space of the humanities, while at the same time, it allows ecocritics to reveal the historical and ideological (mis-) appropriations of nature as a justification for systems of cultural and social oppression.

There is a growing recognition among ecocritics today about nature and cultural diversities in the competitive world dictated by modern technology and changing lifestyles. By concerning themselves with pressing issues of contemporary critical ecological conditions like global warming, climate change, and pandemics, it becomes important for us to accept that nature is a 'space of difference' outside hegemonic discourse. 'It signifies all that is non-civilized'. It is a label for those places that man perceives as being qualitatively apart from "civilized" places. It is the "Other" ... it becomes 'a passive habitat, meaningless in and of itself: it means the trees, the ocean, the mountains, the sunlight, etc., but never has a meaning of its own. In Western thinking, nature is conceived as existing to serve human designs and becomes objectified in the process.

> This human-centered or anthropocentric worldview also led to the development of 'a mechanistic Cartesian metaphysics which sees nature as a dead, inert machine, insensitive to abuse and exploitation by humans' ... In this can be said to lie the roots of the domination and

> exploitation of the natural environment that is at the center of the current global environmental crisis.
>
> (Mutekwa and Musanga, 240)

In his essay "The Climate of History," Dipesh Chakrabarty talks of the colonization of nature by humans where they become 'geological agents' responsible for 'changing the most basic physical processes of the Earth' (quoted in Ghosh, 12). Furthering Chakrabarty's argument Amitav Ghosh says that climate crisis is a crisis of culture and as a corollary is seen as a crisis of imagination. Culture generates in people the desire for comfort and entertainment, for vehicles and appliances which are major sources of carbon emission. A speedy vehicle or a holiday in nature stimulates our imagination generating a desire to possess and control the passive 'other' which is akin to colonizing nature. Postcolonial theory while articulating concepts of alterity and double-marginalization explores the paradoxical closeness between colonizers and colonized for a productive capitalist vision of nature represented in desire. The artefacts, commodities, and pleasures that are imagined by these desires are at 'once expressions and concealments of the cultural matrix that brought them into being' (13). For example, to quote Ghosh, speeding in a car in the forest roads 'excites us because it evokes an image of a road narrowing through a pristine landscape; we think of freedom and the wind in our hair; we envision ... racing towards the horizon' (13). Our capitalist yearnings romanticize nature and seek to colonize it exercising a particular style of power to dominate, restructure, and exert authority over pristine nature which is the silenced other. The resilience of nature has been normative, as depicted in almost all Indian English fiction, and any resistance and retaliation in the form of cyclones, tornados, earthquakes, tsunami, etc. are considered freak events and savage monstrosities.

Starting from Bankim Chandra's *Rajmohan's Wife*, the first Indian English novel published in the nineteenth century, to the recent notable novel, *The Ministry of Utmost Happiness* (2017) written by Arundhati Roy, we find nature positioned and portrayed with a neutral brush. Humans occupy the central position and conceal their unstable position of origin as regards nature. It is worth a mention that both Roy and Ghosh address the problematics of nature and exploitation of it in their non-fictional work rather than evoking the jarring problems of oppression of nature in their fictional narratives. Even in ancient Greek comedies, much like our *Panchatantra*, the non-human world is represented through animals who serve to express that they are important because they make contributions to human culture and not because they represent forces of nature. Non-human nature was anthropomorphized and humans dressed as animals performed and danced on the stage. Animals were used to represent forces of fertility and procreation, thus gaining currency in local mythology and religion.

Literary imagination has signed off nature as being commonplace and tame, only relevant when assisting characters in their sojourn and plot development. It is obvious from the poetry, prose, fiction, art, theatre, and architecture, which are the various modes of cultural activities, that Indian culture is intimately linked with imperialism and Western power policies that have shaped it. Indian writers who shape contemporary culture evade the gnawing issues of climate change. Ghosh critiques this lamentable facet of literary production:

> In a substantially altered world, when sea-level rise has swallowed the Sundarbans and made cities like Kolkata, New York, and Bangkok uninhabitable when readers and museum goers turn to the art and literature of our time, will they not look, first and most urgently, for traces and portents of the altered world of their inheritance? And when they fail to find them, what should they – what can they – do other than to conclude that ours was a time when most forms of art and literature were drawn into the modes of concealment that prevented people from recognizing the realities of their plight?
>
> (14–15)

Concealment and evasion of the plight of the silenced Other is in contradiction to postcolonial ideology.

David Arnold and Ramachandra Guha recognize that the ecological and cultural diversity of the Indian subcontinent was way more diverse and problematic than that of America and Europe. In spite of India being an agrarian country, its economy was and is dictated by the vastly diverse ecology of the subcontinent. They contend that the caste system prevalent in India has grown out of a culture dependent on the environment for sustenance. David Arnold and Ramachandra Guha in *Nature, Culture, Imperialism: Essays on the Environmental History of South Asia* note that

> the multiplicity of endogamous groups constituting the caste system are sharply differentiated according to occupation and ritual status, but it is only now emerging that they might be further distinguished, in many cases, by highly specific relationships with the natural environment. Field data from western and southern India exemplifies a system where endogamous groups (or *jatis*) within a village often had exclusive access to a particular species, resource, or territory, with these individual 'niches' usually having a limited overlap. In some cases this system was elaborated to the extent of two *jatis* of basket-weavers having exclusive control over different plant species for use as raw material—a situation made possible only by the extraordinarily high biological diversity of their surroundings.
>
> (Arnold and Guha, 1995, 11)

Arnold and Guha note that South Asia is basically constituted of settled villages where both the ecological diversity and community culture are preserved by the 'numerous and widely dispersed communities of hunter-gatherers, cultivators, nomadic pastoralists and fisher folk [who] have all contributed' (11) towards this harmonious coexistence.

It is important here to discuss European colonialism and the introduction of scientific knowledge in agriculture in India which entailed a rapid and significant modification of the natural environment, which in turn had profound consequences for life in the cities and the countryside. It is relevant in this context to revisit Jonathan Bate's comment that the 'relationship between nature and culture is the key intellectual problem of the twenty-first century. Clear and critical thinking of the problem will be crucial to humankind's future in the age of biotechnology' (Coupe xvii). Bate's comment makes one realize that European colonization was a prime agent of environmental and social change. Crosby, a leading American scholar in the field of environmental history, in his book *Ecological Imperialism* argues that

> the main reason for the Europeans' success in conquering and colonizing the New World was not, as has been commonly supposed, their superior weaponry, but rather their 'portmanteau biota'—the complex of diseases, plants, and animals which accompanied them and which devastated indigenous cultures and their supportive ecosystems. This ecological invasion ... then paved the way for the creation of prosperous colonial settlements founded on European-style agriculture and stock-raising in zones which Crosby describes as 'neo-Europes.'
>
> (Arnold, 14–15)

As a case in point, I would like to discuss Pankaj Sekhsaria's depiction of the Andaman Islands in his novel, *The Last Wave: An Island Novel* (2014), which is a typical example of the intrusion of human culture into nature whereby the pristine nature is corroded and indigenous people are corrupted. *The Last Wave* is about the life of the people living in the Andaman Islands, an Indian post-colony in the Bay of Bengal. Settlers from the Indian mainland and the descendants from prisoners of the Andaman Jail have cleared forests and acquired aboriginal land in the Andaman Islands. Their interface with the aquatic life, the rainforest, and the indigenous people of the islands, like the Great Andamanese, the Jarawas, and others, is indeed a matter of concern.

> In colonial descriptions, pastoralists [here the indigenous people] were objects of contempt. They were inevitably represented as lazy, improvident, "wretched" as cultivators, lawless, wild, and even mean and cowardly. They were associated with all that was considered evil, ugly and miserable.
>
> (Arnold and Guha, 70–71)

In the appendix of the novel, Sekhsaria gives a detailed chronology of the colonization of the Andaman Islands. In 1789, the British East India Company decided 'to set up colony in the Andaman Islands; Lt Hyde Colebrooke visits the islands, meets native islanders and records some of their language ... the colony that was set up was abandoned a few years later' (283) and after the Sepoy Mutiny of 1857, the British sent the mutineers and criminals to the islands as a penal settlement, known commonly as Kalapani or deportation to the dark waters and later in 1896 the British built the infamous Cellular Jail (283). Occupation of land in the islands began when the prisoners were released and some land was given to them to live in. These settlers invade the unspoiled and until then untouched nature with their culture pushing the forest back. The indigenous Jarawas, an endangered tribe of the Andamans, come in contact with the settlers and their 'portmanteau biota' and contact diseases that they never knew before. Their nakedness is exhibited and sold; they become addicted to tobacco, paan masala, and alcohol and under the influence of the high culture of the settlers, become self-destructive. The novel concerns itself with the nature–culture interface in the Andaman Islands, which in turn involves the existence and survival of the Andaman Rainforest and aquatic life in and around the islands. Sekhsaria points out that the Jarawas are as important to the history of humankind as the Rain Forest, whose survival is important for the survival of humans on earth. David Bhaskaran, a researcher in the novel, is keen on studying crocodiles and other aquatic animals unique to these islands. He discovers that poachers from the island itself and some from Burma and Indonesia visit the crocodile habitats regularly and poach them for their precious leather.

Lynn White, Jr. in her essay "The Historical Roots of Our Ecological Crisis" discusses Aldous Huxley's discourse on 'Man's unnatural treatment of nature and it's sad results' (1203).

> To illustrate his point he[Huxley] told how, during the previous summer, he had returned to a little valley in England where he had spent many happy months as a child. Once it had been composed of delightful grassy glades; now it was becoming overgrown with unsightly bushes because the rabbits that formerly kept such growth under control had largely succumbed to a disease, myxomatosis that was deliberately introduced by the local farmers to reduce the rabbits' destruction of crops.
>
> White, 1967, 1203)

Pankaj Sekhsaria in the chapter "On the Wings of Silence," *The Last Wave* tells us a similar story that the Great Andamanese, who possessed ancient knowledge, contracted a disease from the British and were almost wiped out.

> contact with the British and disease, initiated the process of their almost complete extermination in the period leading up to the Second World

War. Estimated at anywhere between five and eight thousand people when the British set up there colony in the islands in 1858, the Great Andamanese were left with a population of under a hundred individuals when the Japanese came visiting a less than a century later.

(Sekhsaria, 2014, 126)

Sekhsaria explains the vulnerability of the Jarawas to diseases like measles: 'When small, nomadic, forest-dwelling populations come in contact with sedentary, high-density populations, there is always the chance that they will be infected with diseases that might be common in the settled communities' (127).

Human intervention does the same to ecologies. Lynn White commenting on changing ecologies says that 'In many regions terracing or irrigation, overgrazing, the cutting of forests by Romans to build ships to fight Carthaginians or by Crusaders to solve the logistics problems of their expeditions, have profoundly changed some ecologies' (1203–1204). She further argues that our 'ecologic crisis is the product of an emerging, entirely novel, democratic culture' (1205) which was not acute in the cultures of antiquity where there was a pre-Western, pre-Christian culture of animism. 'Christianity, in absolute contrast to ancient paganism and Asia's religion ... not only established a dualism of man and nature but also insisted that it is God's will that man exploit nature for his proper ends' (1208). White believes that:

At the level of the common people, this worked out in an interesting way. In antiquity every tree, every spring, every stream, every hill had its own genius loci, its guardian spirit. These spirits were accessible to men, but were very unlike men; centaurs, fauns, and mermaids show their ambivalence. Before one cut a tree, mined a mountain or dammed a brook, it was important to placate the spirit in charge of that particular situation and to keep it placated. By destroying pagan animism, Christianity made it possible to exploit nature in a mood of indifference to the feelings of natural objects.

(1208)

The narrative of *The Last Wave* depicts a similar situation in Andaman's colonial history of capitalist exploitation by the British:

the Great Indian Railway project on the mainland and the demands of the British Navy meant that timber extraction in the rich forests on the islands grew rapidly and a need arose to induct new labour and expertise into the Forest Department ... The Church was forever willing to oblige the empire, and it soon became the recruiting ground for those to be brought to the islands with promises of plenty and prosperity.

(21)

Thus, both the Church and the Empire are responsible for the destruction of the forests and changing the ecology of the islands. Lynn White's statement that 'As we enter the last third of the twentieth century, however, concern for the problem of ecologic backlash is mounting feverishly' (4) sounds prophetic in the case of the changing face of Andaman rainforests and an anxious world in the vicious grip of the Covid-19 pandemic.

The Last Wave projects a botanist, SK, who expresses his concern over the condition of the once resplendent Andaman Rainforest. He says:

> If you drive beyond the mud volcano in Baratang, you pass through a badly degraded natural forest interspersed with row after endless row of exotic teak, flimsy stands that extend as far as the eye can see ... The brown of the trees, particularly in the dry season, seem to merge with the brown of the soil that was once rich with humus and life. Now it is completely eroded, dead, and sterile ... That part of Baratang is like a furnace, and the villagers there have a serious water problem. It's not just the forests that suffer, people do too – the entire hydrology of the area changes.
>
> (189)

This simply exemplifies that the pace of ecological change has accelerated still further in recent decades, under the auspices of state-directed economic development. While humans tamper with nature, nature has its own means of retaliation proving the insignificance of human ego and prowess. Justifying the title of his novel, *The Last Wave*, Sekhsaria winds up the narrative with a crucial scene where the sea unleashes its anger on humans who are incapable of protecting either themselves or their possessions in the face of the tsunami. The protagonists, Harish and Seema, see:

> The wall of water, even as it kept building up, started to move – towards them. He thought he heard an angry hissing. It was an irate, petulant sea that was coming back ... A huge, solid mass of grey water came rushing in, engulfing the forest camp, lifting it, and then ripping apart the fragile construction as if it were a house of cards. ... The water gushed deep into the forest beyond ... then it hissed viciously as it turned back, withdrawing with a force and a vengeance that made the incoming wave appear benign in comparison. There was a flurry of action, of sound and movement. The swirling of the waters, the roaring of the winds, and the most frightening of them all, the snapping.
>
> (261–263)

Harish and Seema climb to the top of a concrete building, cling to the solid pillars, and try to save themselves. But 'The big, grey-brown mass of water thundered in with uncompromising power, scooping up Harish and Seema and sending them swirling into the vast, now empty, waterscape beyond' (265). The marriage of science and technology, which conceived of nature as a passive habitat to be exploited by humans for their cultural advancement, changed the essence of the impact of humans on the environment of which tsunamis could be a result.

With the advent of firearms and canons in the fourteenth century, men rushed into the forests for potash, sulphur, iron ore, and charcoal, affecting the ecological health of the earth with erosion and deforestation. 'Hydrogen bombs,' says White, 'are of a different order: a war fought with them might alter the genetics of all life on this planet' (1204). Yet humans are heedless of the impending danger of annihilation.

It is relevant to discuss the poem "Of Mountains" to show the degree to which the destruction of nature and the ecosystem is linked to cultural progress. The Mountain, personified in the poem, tells us that in antiquity it had protected humans from the wrath of nature and wild beasts and today man is paying back by blasting it mercilessly for his own economic progress. The Mountain laments:

> When you blast the heart of mountains
> With dynamites and machines
> Do you know how it feels?
> Do you ask even, how it feels?
> Mountains don't speak.
> How can you tell! How can you know?
> Mountains don't shriek, and
> Cannot fight back.
>
> (Das 2016: 27)[1]

The poem points to the greed and insensitivity of humans towards nature and also posits that conserving nature alongside human development presents unique challenges. With technological and scientific progress and its rapid advancement, both cultural and ecologic diversity are under greater pressure, often exacerbated by our desire to manage and control nature. In this context, Gandhi's environmentalism cannot be ignored. It has its roots in his concept of Swaraj. Gandhi had a deep antipathy to urban civilization and a belief in self-sufficiency, self-abnegation, and denial rather than extravagance. Gandhi did not directly advocate going back to nature but to the village and the indigenous sustainable knowledge embedded in rural culture and to rural asceticism and harmony as against urbanization and industrial strife. Finally, we can say that any hope for saving nature and ecological diversity is predicated on human effort to appreciate nature and to understand and respect cultural diversity for sustainability on earth.

Note

1 This poem is written and published by the author of this chapter. See Das, Shruti. "From Ecosensibility to Ecofeminism." *Re-Thinking Environment: Literature, Ethics and Praxis*. Ed. Shruti Das. Authors Press, 2017. 49–59.

Works Cited

Armbruster, Karla and Kathleen R. Wallace. Eds. *Beyond Nature Writing: Expanding the Boundaries of ecocriticism*. University of Virginia Press, 2001.

Arnold, David and Ramachandra Guha. Eds. *Nature, Culture, Imperialism: Essays on the Environmental History of South Asia*. Oxford University Press, 1995.

Coupe, Laurence. Ed. *Green Studies Reader*. Routledge, 2000.

Cronon, William. "Introduction: In Search of Nature." *Uncommon Ground: Toward Reinventing Nature*. Ed. William Cronon. W. W. Norton, 1995. 25–26.

———. "Of Mountains." *The Widening Gyre: An Anthology of Poetry, Prose and One-Act Plays*. Eds. R.N. Panda and S. Das. Oxford University Press, 2016. 27–29.

Ghosh, Amitav. *The Great Derangement: Climate Change and the Unthinkable*. Penguin Books, 2016.

Kerridge, Richard and Neil Shammells. Eds. *Writing the Environment: Ecocriticism and Literature*. Zed Books, 1998.

Mutekwa, Anias and Terrence Musanga. "Subalternizing and Reclaiming Ecocentric Environmental Discourses in Zimbabwean Literature: (Re)reading Doris Lessing's *The Grass is Singing* and Chenjerai Hove's *Ancestors*." *Interdisciplinary Studies in Literature and Environment*. 20.2 (Spring, 2013): 239–257.

Phillips, Dana. *The Truth of Ecology: Nature, Culture and Literature in America*. Oxford University Press, 2003.

Sekhsaria, Pankaj. *The Last Wave: An Island Novel*. Harper Collins, 2014.

Serres, Michel. *The Natural Contract*. Trans. Elizabeth McArthur and William Paulson. The University of Michigan Press, 1995.

White, Lynn. "The Historical Roots of our Ecological Crisis." *Science*. 155.3767 (10 Mar 1967): 1203–1207. DOI: 10.1126/science.155.3767.1203. https://science.sciencemag.org/content/155/3767/1203. Accessed 19.12.2020

Wordsworth, William. "Lines Composed a Few Miles above Tintern Abbey, On Revisiting the Banks of the Wye during a Tour." *Poetry Foundation*, July 13, 1798. https://www.poetryfoundation.org/poems/45527/lines-composed-a-few-miles-above-tintern-abbey-on-revisiting-the-banks-of-the-wye-during-a-tour-july-13-1798. Accessed 19.12.2020.

3 Ecocriticism and Comics

Pinaki De

Both ecocriticism and comics are what Michel Serres and Bruno Latour designate as *quasi-objects* – hybrid combinations assembled in the middle ground between well-established polarities: nature and culture, image and text. For philosopher Michel Serres, a *quasi-object* is something that predates the subject–object distinction. It is neither an active subject nor a passive object, but instead the ground for both of them. It creates a network around itself that makes agency and structure possible. The very existence of *quasi-object* depends on the things around itself; it is itself nothing more than a node of these relations. In *We Have Never Been Modern* (1991, translated later in 1993), Bruno Latour states that our sciences overstate the subject–object and nature–culture dichotomies, whereas, in actuality, a phenomenon often crosses these lines. The *quasi-object* equips us to develop a new model of knowledge that goes beyond simple dichotomies. Latour emphasizes the power of a quasi-object to act as a creative mediator, functioning between terms to produce something new even as it also (re)presents its constituents – the hybrid's transforming ability to make us see or hear or think about something we have not before (Latour, 77–78). Latour further notes: 'In this way, the middle was simultaneously maintained and abolished, recognized and denied, specified and silenced' (Latour, 78). This erasure of boundary markers between dialectical poles ensures that nature and culture are not bracketed separately because 'cultures—different or universal—do not exist, any more than Nature does. There are only nature-cultures, and these offer the only possible basis for comparison' (Latour, 104). Broadly sketching the outlines of ecocriticism as 'the study of the relationship between literature and the physical environment,' Cheryll Glotfelty envisioned a blend of literary criticism and scientific theory, through an interdisciplinary approach, which can 'negotiate between the human and the nonhuman' while exploring the nuances of their intersection as represented in literature (Glotfelty and Fromm xix). Comics are likewise a shifting combination of text and image. According to Scott McCloud, 'the mixing of words and pictures is more alchemy than science' – another kind of assemblage that relies on the tensions and affinities between its constituent arts to animate its content (McCloud, 161).

DOI: 10.4324/9781003383970-5

In the present times, accelerating global warming demands corporate environmental accountability and dwindling non-renewable resources, ecocriticism also needs to consider how some narratives reignite our environmental imagination, providing us with creative scenarios in which we actively rise to meet those challenges. Comics with their intricate polysemic and heteroglossic synergies of image and text invite the readers to become active participants in the narratives they project. They provide 'both visual and verbal interpretive skills' to connect multiple images and bits of text into a comprehensive whole (Eisner, 2). Comics rely not on photorealism but on drawn images, to suggest not 'a direct representation of the world,' but rather 'an interpretation or transformation of the world, with aspects that are exaggerated, adapted, or invented' (Wolk, 20). Tim Morton, writing specifically about environmental aesthetics, proposes that 'art forms have something to tell us about the environment because they can make us question reality' (Morton, 8). Comics as a unique hybrid form literally enhances our perception and makes us pay attention to various micro-narratives that we let pass in our day-to-day experience. The compelling urgency that flows out of graphic images set in motion by 'the dialectic between word and image' provides a popular crossover site for active engagement.

A case in point is the story "A Billion Conscious Acts," included in the fifth collected volume of writer/artist Paul Chadwick's long-running series, *Concrete*. The eponymous central character (formerly Ron Lithgow, a speechwriter) is a normal man whose brain is transplanted into a large, stone body by aliens, and who lives an extraordinary life on earth following his escape. Concrete metamorphoses into a wanderer, whose alien-enhanced eyesight compensates for other sensory disabilities. Concrete has no senses of taste, touch, or smell – along with great strength and endurance, his acute vision is his most valued almost-super power. Dislocated from his original human body, he sees everything anew and the readers are invited to do so as well. In the story, the vagaries of environmental activism are explored when Concrete unwittingly becomes associated with a band of eco-radicals whom he wishes to help after he witnesses the unscrupulous practices of the logging industry firsthand. However, he's also wary of the group's legally dubious methods. 'Perspective is everything,' Concrete advises us elsewhere, because 'we make decisions based on what we see around us' (Chadwick, 9). Chadwick deliberately shifts perspective, from above and below, from within and without, from the local viewpoint to the global perspective in "A Billion Conscious Acts." The reader *sees* the graphic environment in two different ways, as a participant and as a spectator, as a contributor to its ecological state of being, and as the beneficiary (or victim) of those actions. Using the power of image and text, he renders our consciousness a springboard for rethinking our relationships with the environment, and suggests that while saving the rainforests will not be a walk in the park, 'it will be worth it' for the environment's sake and for our own (Chadwick, 156).

Samuel Taylor Coleridge's classic poem about a man trapped within an ocean amidst a great journey, unable to obtain peace, charts a poignant metamorphosis in an updated sequential graphic narrative by Nick Hayes. Hayes transforms the naturalistic ideas of *The Rime of the Ancient Mariner* (1798) in *The Rime of the Modern Mariner* (2012) where the reader experiences an eco-fable set in our current period. In a time, rife with terms such as 'global warming,' 'climate change,' and 'carbon footprint,' the journey from Coleridge's to Hayes's pen seems like a natural and evolutionary growth. In Hayes's *Rime*, the mariner accosts an office worker in a park as he eats his lunchtime sandwich, having casually and ironically discarded the plastic box in which he bought it. The tale he tells is one of the absolute environmental disasters. Stranded in the North Pacific Gyre – a swirling and poisonous whirlpool of plastic waste – the mariner comes face to face with the consequences of our unthinking consumption. After killing the albatross, Hayes's mariner sees all kinds of horrors – a North Pacific drilling barge leaking a 'glossy thick petroleum slick,' swaths of polystyrene bob in the heart of the North Pacific Gyre, and nylon netting in the body of the albatross itself – all rendered in precise but nightmarish line art. Through a lavish dream sequence at the end, the mariner comes to an understanding of his place in nature, a sort of rebirth that has him feeling truly interconnected with life on earth. Back on land, cradled by a bower of ancient trees, his pillow a clump of sweet cicely, the mariner finally understands that the earth, too, has a heartbeat and that we ignore it at our peril. From here on in he vows to remind everyone he meets of this simple truth. The new perspectives provided in *The Rime of the Modern Mariner* enable readers to (re)experience Coleridge's text and face a world that is 'detached from consequence.'

Another sequential graphic narrative that alludes to the infamous Great Pacific Garbage Patch is *I'm Not a Plastic Bag* by Rachel Hope Allison. The work is a wordless, whimsical flight of fancy whose environmental message is far from explicit. It takes the Garbage Patch not as a setting or an environmental problem but as a character in itself. Like Frankenstein's monster, the refuse coalesces into a character with features we can recognize as eyes, a mouth, and hands. The touching narrative gets quite appropriately grim, with this garbage-island-creature discovering that despite its best intentions, it is a blight on the environment. Grotesque yet oddly endearing, the patch goes looking for a friend but repels everything it comes into contact with. The book is filled with clear-eyed, unsentimental depictions of trash, reminding us that a staggering amount of our waste ends up in the ocean. A closing section provides background on the garbage patch and outlines simple steps young readers can take to reduce their own waste. The sequence of images without any accompanying text creates what Douglas Wolk refers to as the 'immersive experience of comics.' Here, the readers of comics actively enjoy 'filling in *all* the blank spaces beyond each panel' – a kind of readerly way, finding that alternatively maps a strange sense of empathy and foreboding onto the narrative (Wolk, 132).

Due to this immersive quality, comics are also a great form that can educate people about intricate issues pertaining to the politics around climate change. French cartoonist Philippe Squarzoni's *Climate Changed: A Personal Journey through the Science* is a case in point. In a scene from the book, the protagonist explains our delicate climate situation to a friend over lunch: 'We pass a certain threshold, and one of the processes determining climate could "snap,"' he tells her. An image of a plate toppling over a table's edge returns again and again throughout the book. This visual code is more effective than a thousand words as it effectively impacts our understanding of the precarious situation in a simple metaphor taken directly from our daily life. The hefty primer, weighing in at 467 pages (not including bibliography, sources, and an index), is a sort of memoir, tracing Squarzoni's own process as he learns about climate science and politics. Along the way, he weaves in personal moments with his life partner, his dog, his travels, and his favourite movies. As meticulously researched as it is illustrated, the comics is like a crash course on science, explaining everything from atmospheric science to the Intergovernmental Panel on Climate Change, from climate disasters to renewable energies, how the emissions in our atmosphere contribute to warming to the benefits and pitfalls of our renewable energy options, adding in his reflections and ruminations about it all. The book's focus on science makes it read at times like a beautifully illustrated textbook. But the book's size is a testament to the sheer complexity of the problem. Squarzoni is interested in political discourse, but he's not taking sides. He doesn't need to as he is not an environmentalist or a scientist or a policy advocate.

> Climate change is an enormous problem; it supplants all the others. Everyone has heard about it, but we don't know what it means. And there are all different kinds of solutions that all mean something different politically. Certain solutions are dead ends, certain solutions drive a vision for society that I don't agree with, and certain solutions drive a vision for society that I do agree with, but that still causes climate change. That's what I wanted to explore.[1]

Squarzoni explores the troubling political and economic implications of rising global temperatures, and what it means, on an individual level, to live in a society that encourages us to consume beyond the limits of earth's resources. It also investigates how the social dimensions of climate change go beyond biophysical impacts and relate to the social and structural factors underlying vulnerability.

It is surprising to know that Sarbajit Sen's *Carbon Chronicles* (*Otho Carbon Kotha* in Bengali) was written two years earlier than Squarzoni's tome and tackles the same problem through a cautionary allegorical fable. The science and politics of 'carbon footprint' is told through the eyes of an impoverished common Indian, Raghu, who is taken for a literal and metaphorical ride by GOD, a capitalist. There are ironic references to Dante's *Divine Commedia* as the GOD unravels the details of various worlds to Raghu and revels in the way 'development' sucks the life out of the earth's biosphere.

The word 'development' is crucial in the Indian context as it somehow sweeps under the carpet any legitimate concern with thorny environmental issues. No book makes it more explicit than *River of Stories* by Orijit Sen. A touching tale woven around the Narmada Bachao Andolan, *River of Stories* by Orijit Sen is widely known as India's first book-length work in non-fictional comics. The story begins with a dream sequence set in the drawing-room of the protagonist, Vishnu, who is a journalist who is on his first assignment. The prologue induces nostalgia with its art and satire, building an ambience that is highly reminiscent of the 1990s. Mostly comical in its tone, the prologue feels linear and crisp and most importantly it effectively introduces the reader to the insensitivity of political leaders towards the Adivasis, which is the central theme of the graphic novel. This, however, is introduced only as a narrative thread and becomes evident only on reading the novel completely. In addition to this, there's a humorous allusion to how people with a humanities background are often seen condescendingly by people in power, especially concerning their stance on sociological and ecological issues. Through a semi-realistic style, the reader is also introduced to Malgu Gayan, whose story runs parallel with the main storyline of the comics. Largely inspired by tribal tales, the readers are ushered to the mystical tale of the creator of the world, Kujum Chantu, through the mellifluous voice of Malgu Gayan who is singing while playing a *rangai* (a string-based musical instrument). The narrative is then intercut with the stories of both – Vishnu and Malgu Gayan. As a part of his research, Vishnu talks to Relku, his domestic help, who belongs to Ballarpur and is an Adivasi herself. A major chunk of the plot about the Adivasis is then explored through Relku's story. The reader gets to witness the plight of the Adivasis at the epicentre of which is insensitivity from the government and its officials. Post-Relku's narration, Vishnu heads to Ballarpur to cover the Andolan. Vishnu's visit to Ballarpur covers the Rewa Sagar Andolan, introducing the reader to a more contemporary scene. Later the reader is also introduced to the article written by Vishnu and the public's reaction to it. The comics end with a comment by the author where he cogently comments (using all the narrative threads and symbolism) on the issue of building a dam on the river. The symbolism of the river is used throughout the work, to represent an ecosystem, the people dependent on it and their wishes, and inter-subjectivity. The river is the focal point of everything in the plot, and the flux of narrative threads is handled brilliantly through the same.

Comics are 'full of enticing blank spaces, in both space and time, for readers to decorate ... but what they look and feel like when we flesh them out isn't the same way we perceive our own environments' (Wolk, 133). According to the noted French philosopher Jacques Rancière, 'Images, are operations: relations between a whole and parts; between a visibility and a power of signification and affect associated with it; between expectations and what happens to meet them' (Rancière, 3). Images and text sometimes produce moments of destabilizing dissemblance and dissonance, engaging the viewer in a suddenly unfamiliar experience that opens the door for a productive (re)imagining of relationships, a suggestion of alternative futures. Comics attempt a more nuanced negotiation of the interstices between the image and the text, exploring the genre's capacity for transgressing its apparent limitations. These texts embrace the immanent ambiguity necessary to produce an aesthetic experience that promotes the play of disparate identities, active self-reflexivity, and heteroglossic dissensus – an experience where politics might flourish. This is in sharp contrast to globalization's long-reach producing monologic ecological narratives to suit its own purpose that gives an appearance of an illusory unity in the rush to represent a world consensus.

Note

1 Squarzoni, Phillipe. "This graphic novelist tells the true story of climate change." Interview by Sara Bernard. grist.org, Oct 17, 2014, https://grist.org/climate-energy/this-graphic-novelist-tells-the-true-story-of-climate-change/

Works Cited

Allison, Rachel Hope. *I am Not a Plastic Bag*. BOOM! – Archaia, 2013.
Chadwick, Paul. *The Complete Concrete*. Dark Horse Comics, 1994.
Eisner, Will. *Comics and Sequential Art: Principles and Practices from the Legendary Cartoonist*. W. W. Norton & Company, 2008.
Glotfelty, Cheryll and Harold Fromm. *The Ecocriticism Reader: Landmarks in Literary Ecology*. University of Georgia Press, 1996.
Hayes, Nick. *The Rime of the Modern Mariner*. Viking, 2012.
Latour, Bruno. *We Have Never Been Modern*. Trans. Catherine Porter. Harvard University Press, 1993.
McCloud, Scott. *Understanding Comics: The Invisible Art*. William Morrow Paperbacks, 1994.
Morton, Timothy. *The Ecological Thought*. Harvard University Press, 2012.
Rancière, Jacques. *The Future of the Image*. Trans. Gregory Elliott. Verso, 2009.
Sen, Orijit. *The River of Stories*. Kalpavriksh, 1994.
Sen, Sarbajit. *Otho Carbon Kotha*. Nagarik Mancha Kolkata, 2012.
Squarzoni, Phillipe. *Climate Changed: A Personal Journey through the Science*. Abrams & Chronicle Books, 2014.
Wolk, Douglas. *Reading Comics: How Graphic Novels Work and What They Mean*. Da Capo Press, 2008.

4 Dialectics of Environment through the Prism of Fiction

An Overview of Amitav Ghosh's *The Hungry Tide*

Sujit Malick

I

The word 'dialectics,' as a concept, is often associated with the Marxist School of philosophy but etymologically it denotes a systematic method of argument that attempts to resolve the contradictions in opposing views. In this chapter, the word 'dialectics' hardly bears the same meaning, which connotes usually a deeper level of understanding. Amitav Ghosh in *The Hungry Tide* depicts the incidents of Morichjhapi massacre that occurred during the tenure of the Left Front Government. The episode is presented through the diary of Nirmal, one of the major characters of the novel. Morichjhapi was selected as a location for Project Tiger[1] by the World Wildlife Foundation for environmental concern and was framed by the experts of the world, but the state government took stern measures to vacate the land, resulting in bloodshed and death. Hence the present chapter would show how the international environmental policy backfires/contradicts the interest of indigenous lives of Sundarbans, viewed as dialects of environment.

II

The Rio-Earth Summit (1992) initiated a unique way of saving the blue planet. In the same year, Rachel Carson[2] in her book *Silent Spring* cautioned us to reconceptualize the existing method of global conservation and create an awareness mechanism for the world. United Nations Development Programme, by extending its branch, reincarnated itself as United Nations Environment Programme (UNEP), and it formulated the framework that would prevent further depletion of the earth, and in anticipation, various forums and blocks of nations, especially the developed nations with their exclusive agenda, rallied behind the UNEP to show the solidarity of formulating effective conservation resolution. Kyoto Protocol (2005),[3] believed to be the death knell of carbon emission, came with certain annexures, framing detailed functions of the stakeholders. It is said that in Annexure I:

DOI: 10.4324/9781003383970-6

> Parties include the industrialized countries that were members of the Organization for Economic Co-operation and Development.
>
> (9)

In the Non-Annex of the United Nations Framework Convention on Climate Change (UNFCCC) in 2005, Kyoto Protocol diluted the role of developing nations, wherein imagined parties would be mostly developing nations without certain bindings. Article I of the UNFCCC designated developed nations to envisage various effective norms, and it also empowered them to execute and allocate funds so that depletion of the ozone layer could be halted. A close reading of the convention visualized how developed nations strategically manipulated the norms and developing nations were only allowed in the fringe area, namely in the Non-Annexure. Therefore, it could be grossly assumed that the developed nations never renounced their vested interests. Moreover, they tried to incorporate laws in favour of their motion, leaving the 'wretched of the earth' perturbed. This chapter attempts to revisit the role of such conventions, thereby exposing the inherent contradictions that lie in the conservational policies as represented in Amitav Ghosh's *The Hungry Tide*.

III

Ghosh, in almost all of his novels, negotiates between culture and climate change. It is quite interesting that, in his non-fiction *The Great Derangement*, Ghosh laments over the crisis resulting from climate change which is a primarily cultural phenomenon. In this connection, a literary review in *The Hindu* observes:

> Ghosh argues that contemporary culture has largely failed to confront climate change, partly because of the historical elision of various modernity in the favour of the monolithic paradigm of European modernity, currently being toted even by supposedly west – sceptic ideologues as those of Hindutva.
>
> (10)

Ghosh rightly visualizes the ensuing calamities due to climatic change looming large on the coastal cities around the world. In *The Sea of Poppies*, Ghosh exposes the Western imperialistic attitude that impacts the ecological balance of the coastal area. It is mentioned in the novel that almost 25,000 acres of fertile land were completely ruined, and the owner of the land, zamindar Neel Rattan Halder, was deported to Mauritius as an indentured labourer. The recurrent motif of environmental degradation in *The Glass Palace* seems to be an unconscious continuation of Ghosh's previous novel where he criticized the petroleum industry of Burma. Interestingly, the vague notion of 'impeccable economic logic' postulated by Larry Summers, former chief of the World Bank, has been subverted here. The very process of decolonization continues, following the exact method of apparent development that their predecessors followed. Here, an important excerpt from Dipesh Chakrabarty can be recalled:

> The period of Great Acceleration (Keeling Curve) is precisely the period of great decolonization in countries that had been dominated by European Imperial Powers.
>
> (102)

An implicit revision of the motif of exploitation of ecological condition appeared again in *The Hungry Tide* where Ghosh described a symbolic journey of Piyali who came to the Sundarbans to study the endangered Irrawaddy dolphin. Her meeting with the local fisherman, Fokir, and her semiotic exchange of communication with him set the mood of the novel. A meta-narrative of the 1970s Bengal politics is introduced by the novelist as Kanai discovers the notebook written by his uncle, Nirmal, a headmaster of Hamilton`s utopia at Lusibari. This hitherto unread notebook narrativizes the account of Morichjhapi, a utopian Dalit resistance in Nirmal's vision as a socialist. Bangladeshi Dalit refugees, being tortured by Muslim communalists and by some upper-class Hindus of Bangladesh, took refuge in India and the government deported them to Dandakaranya of central India, but they abandoned the camp as they lacked the balmy touch of the environment in which they were habituated in their native land. Anup Jalais mentions in his text *Dwelling on Morichjhanpi*:

> The soil was rocky and the environment was nothing like they had ever known. They would not speak the language of that area and local people treated them as intruder attacking them with bow, arrow and other weapons.
>
> (1757)

Around 150,000 people started settling on the island of Morichjhapi under the leadership of Satish Mondal; they installed the basic amenities of livelihood by clearing the forest area. But, the Left Front Government, a former sympathizer of the cause of refugees, now served a caveat to the dwellers of the island to vacate it as the place was to be a sanctuary of wildlife. The people of the island put up a tremendous resistance against the tyranny of the police force that inflicted inhumane torture upon the hapless Dalit relocated people, and as expected, they succumbed to the lethal onslaught of the state machinery. Causalities were tremendous. Unfortunately, no record or case is archived in the government records. Amitav Ghosh resituated the episode of Morichjhapi in the long-preserved notebook of Nirmal.

> He started writing it on the morning of 15 May 1979. In a place called Morichjhapi.
>
> (117)

Ghosh gave a valuable historical insight that might coincide with the lived situation of the island as that of Ross Mallick. Nirmal's account of the situation is potent evidence of the historical moment:

> the refugees may have assumed that they would not face much opposition from the state government. But this was miscalculation: the authorities had declared that Morichjhapi was a protected forest reserve and they proved unbending in their determination to evict the settlers. Over a period of about a year there had been a series of confrontations between settlers and government.
>
> (119)

When they refused to oblige, as Nirmal records, there were atrocities.

> The siege went on for many days and we were powerless to affect the outcome. All we heard were rumours: despite careful rationing, food had run out the settlers had been reduced to eating grass.
>
> (260)

The government desperately tried to uproot the settlement of the unarmed settlers from the island; humanity remained unattended; rather, on the pretext of wildlife protection, the administration unleashed brutal police force to quell the unity of the oppressed class. Their reason:

> The island has to be saved for its trees, it has to be saved for its animals, it is part of the reserve forest, it belongs to the project to save tiger which is paid for by people from all around the world.
>
> (261)

The last part of the excerpt is quite interesting – 'paid for by people from all around the world,' which indicates the aid of foreign funds, obviously donations from developed countries that would conserve the endangered animals like the tiger. The human settlement was sacrificed to save the forest, leaving an ironical undertone. The law of wildlife framed by developed nations indicates several lacunae in its implementation. The UNFCCC made certain allowances on 'environmental refugees' evicted from their settlement because of climatic change as well as political turbulence. Over one lakh settlers of Morichjhapi were dealt with harsh cruelty. The notebook mentions the tragic fate of Morichjhapi in Ghosh's novel. Foreign funds were meant to prevent the depletion of the environment, but they excluded the welfare of human beings. Rather they perished, as the framework of the UNO did not judiciously allocate the provisions for survival and rehabilitation. Notwithstanding the suppressive measure, resistance came from the voice of the Third World and, in this novel Kusum delivers a fiery speech during demonstrations at Morichjhapi:

> Those people are squatters; that land doesn't belong to them; it's government property. How can they just seize it? If they're allowed to remain, people will think every island in the tide country can be seized. What will become of the forest, the environment?
>
> (213)

Piyali, the cetologist from Seattle, disembarks on the bank of Matla in the Sundarbans to study the movements of Irrawaddy dolphins. Piyali, a representative of the developed nations of the Western world, dedicates herself to conserve the endangered marine animal, but her quest proves futile and her study ends abruptly with Fokir's ill-fated death. The beginning of the novel shows how Piyali spots Fokir in the same manner as she usually searches for dolphins:

> the one he [Fokir] had been wearing when she first spotted him with her binocular – had been laid out to dry on the boat's hood.
>
> (72)

The gadget binocular that Piyali uses for tracing dolphins helps her to gather information: later she carefully discharges it to preserve the endangered creatures. Having been brought up in the West, she knows how to survive the onslaught of nature; however, she unconsciously attracts Fokir, who is a representative of the East who eventually succumbs to his death due to the climatic turbulence. Amitav Ghosh triumphs in displaying a meaningful bond between Piyali and Fokir – a relation beyond language and culture. Moreover, the Piyali – Fokir bond organically demonstrates, without any stereotypical designations, the injury of Western environmentalism on indigenous ethical understanding. A critical examination portrays that Fokir, an inhabitant of the tide country, is well acquainted with the pulse of its soil. Ironically, Piyali becomes the very reason for the untold miseries faced by the settlers including Fokir:

> someone standing in for the men who had destroyed Fokir's village, burnt his home and killed his mother; he had become a token for aversion for human beings in which a man such as Fokir counted for nothing, a man whose value was less than that of an animal.
>
> (327)

IV

Conservation policy adopted by the developed nations proved disastrous and, according to Ghosh, should be reconsidered. Otherwise, the rhetoric of conservation to protect the world from climatic change will go astray. In an interview with Shalini Umachandran, Ghosh very clearly articulated his thoughts when he was questioned about *The Hungry Tide*: writing *The Hungry Tide* played a large part. In Sundarbans even around 2000, you could see quite clearly the impact of climatic change (*Times of India*, 2016).

Amitav Ghosh's *oeuvre* is a continuous critique of the Western narrative of environmental conservation, revealing the hypocrisy and indifference towards those people who are reeling under the scathing attack of climatic turbulence. This chapter, therefore, is a humble endeavour to examine the inextricably interwoven text and subtext of *The Hungry Tide*, thereby exposing the futility of the Westernized notion of conservation and how it adversely impacts the lives of the Third World nations.

Notes

1 Project Tiger, administered by the National Tiger Conservation Authority, was a tiger conservation programme launched in April 1973 by the Government of India during Prime Minister Indira Gandhi's tenure.
2 Rachel Carson (1907 - 1964) was an American marine biologist, author, and conservationist who engaged in global environmental movements and the natural history of the sea.
3 The Kyoto Protocol was an international treaty which extended the 1992 United Nations Framework Convention on Climate Change that commits state parties to reduce greenhouse gas emissions.

Works Cited

Ghosh, Amitav and Allen Lane. *The Great Derangement: Climate Change and The Unthinkable*. Chicago: University of Chicago Press, 2016.

——— *The Hungry Tide*. Noida: Harper Collins Publishers India, 2008.

——— "Interview by Salini Uma Chandran." *The Times of India*. New Delhi. 19 July, 2016. print.

Jalais, Anup. "Dwelling on Morichjhanpi: When Tigers became "Citizens", Refugees "Tiger-Food"." *EPW*. 40.17 (Special article, January 2005): 1757–1762. Web. 22 Oct, 2016. www.epw.org.in/show/article php

Kyoto Protocol. "United Nations Framework Convention on Climate Change." Kyoto. 2005. Web < http// unfccc.int>essential _ background > item

5 From Ecocriticism to Omninaturalism
The Green Consciousness and Intercorporeality

Sourav Kumar Nag

I

Introduction

In this discussion, my primary point of contention is that the Western environmentalist philosophies are not adequate to generate ecogenic consciousness. The point of my departure is A. Nichols's idea of 'urbanature' (2011). In his book *Beyond Romantic Ecocriticism: Toward Urbanatural Roosting* (2011), Nichols argues that the new word 'urbanature' suggests

> that nature and urban life are not as distinct as human beings have long supposed. It suggests all human and nonhuman lives, as well as all animate and inanimate objects around those lives, are linked in a complex web of interdependent interrelatedness.
>
> (2011: xiii)

Nichols's formulation of the 'interdependence' of the human and non-human or the living and inanimate is not new, though it certainly revises a body of old environmentalist philosophies. The point of my argument in this chapter is that the idea of mere interdependence between the human and non-human does not adequately formulate man's relationship with nature. In this context, let us move towards a new coinage, a new idea, called 'omninature' that should not be confused with the cliched Pantheistic philosophies. Omninature posits nature as an all-embodying entity. It incorporates the whole of the creation. In short, there is nothing outside nature and we cannot 'return' to nature because we are already, always within it.

Every ecocritical writing begins with a fable – a dream for a sparkling tomorrow that is free from the complex capitalist machinery and sustained by the rule of an Edenic environment where man is posited in perpetual Adamic indolence. Gerrard has rightly commented: 'so the foundling text of modern environmentalism not only begins with a decidedly poetic parable but also relies on the literary genre of pastoral and apocalypse' (2). The most common form of environmental justice, sought by many critics,

DOI: 10.4324/9781003383970-7

is posited in the popular catchphrase 'return to nature.' The recent coronavirus outbreak as the most threatening pandemic after World War II convicted human agency as the only species responsible for the deadly mutation of the virus, as though nature without man would be the perfect setting for the progress and prosperity of the Mother Earth. One must not forget that nature without a man is non-existent and a man without nature is an impossibility, precisely because man embodies nature like any other living organism and non-living entities and at the same is conscious of it. Anything beyond human consciousness is non-existing. The green moral of conventional ecocriticism heightens the image of nature as something transcendental and something beyond human reach:

> Putting something called Nature on a pedestal and admiring it from afar does for the environment what patriarchy does for the figure of Woman. It is a paradoxical act of sadistic admiration. Simone de Beauvoir was one of the first to theorize this transformation of actually existing women into fetish objects.
>
> (Morton, 5)

The point of my argument is that the so-called 'interdependence' between man and nature presupposes a hierarchy – a master/slave power play. Interdependence or interrelatedness does not free the non-human agency from being fetishized, ghettoized, and subdued by man. Man is interdependent with the non-human agencies in an unequal, unproportionate relationship. It is not enough to know that the trees are our friends because they give us oxygen, and other sustenance, that the clouds give us rain and that we produce different resources from the animals. Therefore, the theory of interrelatedness can hardly raise green consciousness and prevent the exploitation of nature.

II

Ecocriticism – Dated?

The ecocritical studies promoted by the Western environmentalists to save nature failed to raise serious concerns because the ethical environmentalism represented nature as something distanced from human society. Man has been made into a spectator whose ecocritical gaze is distant and obsolete. Beginning with the biblical account of the Paradisal happiness enjoyed by Adam and Eve and moving through Rousseau's notion of 'noble savage' down to the conventional ecocritical gaze, we may find nature as something of a giver that gives happiness to man and distancing away from it may cause unhappiness and corruption: 'most of our ills are of our own making ... we might have avoided them all by adhering to the simple, uniform and solitary way of life presented to us by nature' (Rousseau).

Very recently, one may find the shadow of doubt and disillusionment among the environmentalists about the future of ecocriticism. They do not find the goal of ecocriticism heading towards realization outside the air-conditioned conference rooms, hardbound books, and doctoral or postdoctoral theses:

> And the word environment. Such a bloodless word. A flat-footed word with a shrunken heart. A word increasingly disengaged from its association with the natural world. Urban planners, industrialists, economists, developers use it. It's a lost word really. A cold word, mechanistic, suited strangely to the coldness generally felt toward nature.
>
> (Williams, 9)

Nevertheless, new ideas that explore the relationship between nature and man are being offered by environmentalists such as Ashton Nichols, Stacy Alaimo, Timothy Morton, and others. Stacy Alaimo, for example, asserts in *Bodily Natures: Science, Environment, and the Material Self* (2010) the 'trans-corporeality' between the human and 'more-than human nature' (2) to illustrate 'the interchanges and interconnections between various bodily natures' (2). Ashton Nichols in *Beyond Romantic Ecocriticism: Toward Urbanatural Roosting* (2011) revisits the Wordsworthian and Thoreauvian aesthetics to incorporate the environment within the domains of urbanity. He refutes the conventional distinction between nature and culture (xiv). He explains:

> With this in mind, my goal is to move, not only back (to nature) but also forward (toward urbanature), forward into the experience of living in a world where nature and culture are no longer seen as distinct. This new world is one where the natural and the urban mingle—just as they do so easily on any map these days—just as they do now in my computer screen on Google Earth or Map Quest.
>
> (4).

Joy Williams speaks of the human indifference to nature. He condemns the movement of human civilization far away from the non-human world. Like Prometheus, man is confined to the rock of narcissistic self-glorification:

> I DON'T WANT to talk about me, of course, but it seems as though far too much attention has been lavished on you lately – that your greed and vanities and quest for self-fulfilment have been catered to far too much. You just want and want and want. You believe in yourself excessively. You don't believe in Nature anymore. It's too isolated from you. You've

> abstracted it. It's so messy and damaged and sad. Your eyes glaze as you travel life's highway past all the crushed animals and the Big Gulp cups.
> (7)

Williams's words certainly remind one of Wordsworth:

> The world is too much with us; late and soon,
> Getting and spending, we lay waste our powers;—
> Little we see in Nature that is ours.
> ("The World Is Too Much with Us," 1–3)

It is truly comforting to see the Western critics disowning the barriers between nature and culture and moving towards a new dimension of thought that defines human nature and the external nature or that the environment be interconnected. The framework for the emerging trend of man's interconnectedness with nature is not new. In 1980, it was iterated in the first United Nations' World Conservation Strategy:

> Ultimately the behaviour of entire societies towards the biosphere must be transformed if the achievement of conservation objectives is to be assured. A new ethic, embracing plants and animals as well as people, is required for human societies to live in harmony with the natural world on which they depend for survival and well-being. The long term task of environmental education is to foster or reinforce attitudes and behaviour compatible with this new ethic.
> (IUCN report, 1980)

My chapter pushes forward a little beyond what Nichols termed 'Urbanatural' to a new domain of 'omninatural' – a new coinage that should not be confused with Spinoza's Pantheistic doctrine. In pantheism, nature was considered to be an all-emanating force that encircles everything. The word pantheism is a combination of 'pan' (meaning all) and *theos* (meaning 'God'). In pantheism, God is seen as a combination of all the five elements and the real manifestation of God. Nature, therefore, in pantheism is something associated with divinity.

III

From Urbanature to Omninature: Indian Philosophy and Beyond

Omninature is an inclusivist doctrine. It goes beyond the mere claim of interconnectedness, as Nichols argues in his book. Omninature presupposes that nature embodies the whole of creation. Every human or non-human agency, every living or inanimate thing is a part of that intercorporeality. There are religious and philosophical parallels to the concept of omninature. In the Srī-Vaiṣṇava tradition, creation is called the emanation of the divine. In the

Vedanta, Brahman is the One Blissful Reality that is all pervasive. Brahman is infinite and his relationship with the universe is that of love. It is not that the elements of creation are not parts of Brahman but full Reality itself: 'The essence of Vedanta is that there is but one Being that every soul is that Being in full, not part of that Being' (Vivekananda, 51). In the Vedanta, nature is called *prakrti* or the original form of anything. It is the primary substance of both the living and non-living entity. The amazing thing about this philosophical idea is that *prakrti* or nature is not pushed beyond the self (cogito) as something distant, something there to reach. In Western environmental ethics, nature has been viewed as non-human. The non-human world is supposed to depend on the human world for the production of natural and artificial resources through rational deliberation (Soper, 1995: 37–38, quoted in Baindur, 92). Both *Nyāya* and *Sāṃkhya* schools accept such distinction between the human and the non-human or natural world, though the trajectory of the philosophical ideas is quite different from the Western ones. The distinction between the human and the non-human, the knower (*kṣetrajña*) and the subject of knowledge (*kṣetra*) is a relative one. In *Bhagavad Gīta*, the *kṣetrajña* is not the experiencer of knowledge but becomes aware of the process of thinking through the realization that both the mutable and immutable natures belong to The Lord, the Creator. In it, man is not the creator of resources though he can be aware of the Creation. The moment of knowledge is also the moment of a union between *prakrti* (the nature principle) and *puruṣa* (the witness principle) (Baindur, 93). In other words, nature is the Creation, and The Lord is the Creator and man can only be the conscious of it. The shift in the position of man from the creator of his world to the thinking agency makes room for accepting nature as something original, innate, and unchanging. Man (*puruṣa*) is but a part of nature (the Creation or the *prakrti*). He and nature share an intercorporeality that extends beyond the human agency and the world of sensory cognition:

etad-yonīnibhūtāni
sarvāṇītyupadhāraya
ahaṁkṛtsnasyajagataḥ
prabhavaḥ pralayas tathā

(Bhagavadgīta, 7.6)

> ['All created beings have their source in these two natures. Of all that is material and all that is spiritual in this world, know for certain that I am both the origin and the dissolution' (ŚrīlaPrabhupāda, 546)].

The Indus, Vedic, Buddhist, and Jain traditions promote intercorporeality between man and nature. Nanditha Krishna in *Hinduism and Nature* (2017) offers interesting examples of how the traditional Indian lifestyle incorporates good practices with the human societies that keep the environment healthy:

> The Indian housewife starts her day by cleaning the space outside the front door and decorating it with beautiful designs made of rice flour. Apart from beautifying her home, she is also feeding the ants and does not need to spray insecticide to keep them out.
>
> When she bathes, she prays that the water may be as sacred as the River Ganga, which has proven antimicrobial qualities.
>
> She encircles the pipal tree seven times in a ritual binding the Indus, Vedic, Hindu, Buddhist, Jain, and tribal traditions. It is a unique tree that filters impurities in the air and releases oxygen day and night.
>
> She pours water over the sacred basil plant—tulsi—in the centre of her house, for it prevents coughs, colds and fevers.
>
> (9)

These customs have been followed in India for ages. What is interesting about these practices is that they promote the intercorporeality of internal/corporeal and external/terrestrial natures. The *tulsi* or the basil in the courtyard of an Indian house is as important a part of the daily existence of the inmates as they are to each other. In Vedic literature, all of nature was considered a 'divine part of an indivisible life force uniting the world of humans, animals and plants' (Krishna, 11). Nature was not deified as was by the Pantheistic philosophers and the Romantics. *Prakrti* was defined as the basic quality of someone or something. The notion of human culture as distinct from nature is a Western import. Therefore, *prakrti* is the primary substance of anything, and even though man civilizes himself or cultures himself, he cannot move beyond his *prakrti*. The *prakrti* of man is his humanness and he cannot get rid of it because separating humanness from man is to deny his existence. The Sāṃkhya School refers to three different innate qualities or *guna*, such as *rajas* (creative activity), *sattva* (calmness of preservation), and *tamas* (destruction). The perfect balance of these three *gunas* 'is the basis of all observed reality' (Krishna, 12). Therefore, nature is not distinct from reality, be it a factory or a busy road or a museum; nature is everywhere and everything in nature.

In Greek philosophy and literature, the intercorporeality between man and nature is found. In Ovid's *Metamorphoses*, Daphne was turned into a laurel bush when Apollo chased her; Syrinx was turned into reeds to escape advances of Pan. Adonis was given birth to by a myrrh plant and Young Cyparissus turned into the melancholy Cypress tree:

> Scarce had she made her prayer when through her limbs
> A dragging languor spread, her tender bosom
> Was wrapped in thin smooth bark, her slender arms
> Were changed to branches and her hair to leaves;
> Her feet but now so swift were anchored fast
> In numb stiff roots, her face and head became

The crown of a green tree; all that remained
Of Daphne was her shining loveliness.

(*Ovid* Bk.I Lines, 550–557)

Thesmophoria, a festival of ancient Greece, was celebrated in the honour of Demeter and her daughter Persephone, and Phallephoria was celebrated in the honour of Dionysus. The former was associated with harvest and fertility and female agency while the latter with penis, fertility, and male agency. These two festivals represent maleness and femaleness metaphorically and the contribution of both the sexes in creation.

IV

Omninatural Shades: Rabindranath Tagore's Environmental Philosophy

'Omninature' is a concept that promotes green consciousness. By green consciousness, I mean a simple awareness that the nature of man and the nature of the environment are the two parts of the same creation. If a man destroys the environment, he also destroys his own nature unknowingly. This green consciousness is essential for the preservation of nature. Nature has been seen as sacred to man and returning to nature a utopian dream. But omninature posits nature in the orbit of human consciousness as something related to his own self. This can make man conscious that his own nature is a part of the macrocosmic nature. In the following lines from the "Immortality Ode" one can find that Wordsworthian representation of nature is omninatural:

Then sing, ye Birds, sing, sing a joyous song!
And let the young Lambs bound
As to the tabor's sound!
We in thought will join your throng,
Ye that pipe and ye that play,
Ye that through your hearts to-day
Feel the gladness of the May!
What though the radiance which was once so bright
Be now for ever taken from my sight,
Though nothing can bring back the hour
Of splendour in the grass, of glory in the flower;
We will grieve not, rather find
Strength in what remains behind.

(168–180)

The intercorporeality between man and nature is found in abundance in Tagore's works. In Tagore's short story "Bolai," the eponymous protagonist is not merely a lover of nature but has embodied the verdurous symphonies in his soul. The story begins with a philosophical observation of human soul embodying elements of the non-human world:

> In reality we call that human which has blended all the animals within ourselves, combined them into one: penned our tigers and cows into one enclosure, trapped our snakes and mongooses in the same cage. In the same way we give the name raag to that which combines all the *sa-re-ga-ma*s within itself and creates music—after which the notes can no longer make trouble—but still, even within the music an individual note may stand out from the others: in one raag the *ma*, in another the *ga*, in yet another the flat *dha*.
>
> (Tagore, "Bolai")

To Bolai, nature is not just a distant entity to be perceived sensuously but a dynamic flow that can be felt and internalized. When he sees green grass covering a slope all the way, he feels that the grass is a playful rolling mass of green: 'Often he would climb up the slope and roll down himself-his entire body itself become grass-and as he rolled, the blades of grass would tickle the back of his neck, and he would laugh out loud' (Tagore, "Bolai"). Tagore's vision of the intercorporeality of man and nature is further asserted in *The Red Oleander* (1925). In the play, the character of Nandini is a metaphorical embodiment of nature:

> She has her mantle the green joy of the earth. That is our Nandini. In this Yaksha Town there are governors, foremen, headmen, tunnel diggers, scholars like myself; there are policemen, executioners, and undertakers-altogether a beautiful assortment! Only she is out of element.
>
> (554)

Nandini is out of element because she is unlike others. She has learnt her lesson, felt the 'green joy of the earth.' For Tagore, it was the spirit of unity that propagated the complex web of creation on earth. No one is alone, isolated: 'The leather binding and title-page are parts of the book itself; and this world that we perceive through our senses and mind and life's experience is profoundly one with ourselves' (Tagore, *The Religion of Man*, 15). In his ecological cosmopolitanism, he surpassed Wordsworth. He envisioned a society where every living organism and non-living object will be interrelated, inter-embodied. Tagore's environmental philosophies are both spiritual and rational. In *The Religion of Man* (1922), the Indian mystic disseminates his environmental ideas that are replete with the utmost precepts of rationality. His argument behind the inter-embodiment of creation is Darwinian. From the first germs of life on earth to the multicellular life, the unconscious process of creation finds its realization in man. Man has attained the realization of creation in his spirit as well as in his own body.

Man exists when he is encircled by the waves of creation and 'he misses himself when isolated' (15). The discovery of the self is always inter-personal, and his multi-personal humanity is immortal.

He was deeply concerned about the rapid degradation of the environment. A pioneer of progress, Tagore always wanted to maintain harmony between the environment and civilization. In *Tapoban* he praised the coexistence of man and nature in ancient India (*Rabindra Rachanabali*, Vol-7, P-689). In *Muktadhara* (1922), a symbolic play, Tagore has critiqued human greed for material wealth. The English word for 'Muktadhara' would be 'the waterfall.' In the play, the king of Uttarakut decreed to build a dam to confine the waters of the mountain spring for economic well-being. In Santiniketan and Sriniketan, Tagore initiated nature festivals called *Briksharopana*,[1] *Halakarshana*,[2] and *Barshamangal*[3] (Sarkar and Ghosh, 26).

One may object that Tagore's idea of the culmination of the evolution process in man is a humanistic glorification of the human agency that it is fed on a tendency to exult the status of man in the chain of beings. But we must not forget that both the epistemological and ontological precepts are the products of human consciousness. Therefore, the denial of man's involvement in shaping the world seems to be useless. Tagore's idea of the environment is not anthropocentric but humanistic. The realization that man is inter-embodied with the universe, that he is a part of the greater humanity, may raise ecogenic consciousness. It cannot be denied that man with his rationality has shaped his world, but he certainly is not the 'master' of the world. He is only a part of the creation. He may die but his greater humanity will not. Therefore, he is immortal in a way and should present the gift of a better, greener earth to the next generations to come. To conclude I should allude to a few lines by Tagore from *Gitanjali*:

> The same stream of life that runs
> through my veins night and day
> runs through the world and dances
> in rhythmic measures.
>
> ("Stream of Life," 1–4)

Notes

1 *Briksharopana* – On 14 July 1928, Rabindranath Tagore started the Briksharopana Utsav (Tree Planting Ceremony) by planting a bokul sapling in the nearby Pearson Pally. This became a regular festival from 1936 and continues even today. It is performed during the rainy season that helps the plants to grow. It is accompanied by music, dancing, and Vedic chanting as young boys and girls dress in natural ornaments and finery crafted with flowers and leaves.

2 *Halakarshana* – This is the ploughing ceremony that is a symbolic tribute to the activity of ploughing the land. This ceremony takes place on the 23rd day of Shravan or 8/9 August to celebrate the dignity of farming and nurturing of tillable land.

3 *Barshamangal* – This is a soul-stirring creation by Tagore to celebrate and describe monsoon in India. The season's colours brought profound melodious harmony in the poet's mind.

Works Cited

Alaimo, Stacy. *Bodily Natures: Science, Environment, and the Material Self.* Indiana University Press, 2010.

Baindur, Meera. *Nature in Indian Philosophy and Cultural Traditions.* Springer, 2015.

Crawford, Walter Byron. *A Collated Text of the Preface to Lyrical Ballads, 1800–1850: Including the Appendix, 1802–1850: (with Certain Passages Collated with the Advertisement to Lyrical Ballads, 1798).*The Author, 1961.

Krishna, Nanditha. *Hinduism and Nature.* Penguin Books, an Imprint of Penguin Random House, 2017.

Marvell, Andrew. "The Garden." Poetry Foundation. https://www.poetryfoundation.org/poems/44682/the-garden-56d223dec2ced.

Morton, Timothy. *Ecology without Nature Rethinking Environmental Aesthetics.* Harvard University Press, 2009.

Nichols, Ashton. *Beyond Romantic Ecocriticism: Toward Urbanatural Roosting.* Palgrave Macmillan, 2011.

Ovid. *Metamorphoses.* Trans. A.D. Melville. Oxford University Press, 1986.

Prabhupāda, Śrīla. *Bhagavad-gītā As It Is.* ISKCON, 2010.

Rousseau, Jean-Jacques. "A Dissertation on the Origin and Foundation of The Inequality of Mankind and Is It Authorised by Natural Law?" *Rousseau: On the Origin Inequality: First Part*, www.marxists.org/reference/subject/economics/rousseau/inequality/ch01.htm.

Sarkar, Sreemanta, and Chaitali Ghosh. "Environment and Development: A Visit to the World of Tagore." *International Journal of Inclusive Development*, vol. 1, no. 2, Dec. 2015, pp. 23–27.

Tagore, Rabindranath. "Bolai." Trans. Prasenjit Gupta, ISSN 1563-8685. www.parabaas.com.

———.*Gitanjali.* Digireads Com Publishing, 2019. https://parabaas.com/translation/database/translations/stories/ghtml

———. *Red Oleanders.* Rupa & Company, 2002.

———. *The Religion of Man.* George Allen & Unwin Ltd., 1922.

Williams, Joy. *Ill Nature: Rants and Reflections on Humanity and Other Animals: Meditations on Humanity and Other Animals.* Lyons Press. 2001.

6 Exploring Eco-criticism and Eco-feminism

A Re-reading of *Wide Sargasso Sea*

Mohana Das

I

Introduction

Eco-criticism can be defined as a movement that has emerged as part of literary criticism. It is considerably new and has developed in the past three decades to enhance the field of study of human behaviour concerning the non-human environment or nature. The term implies ecological literacy and raises questions about the future of ecology. Ecology has advanced from being a description to advocacy after the 1960s. The stories related to ecology present choices that are ethical and are connected to people and their lands. This research will therefore help in understanding this interdisciplinary field that appeared around the 1990s. Assessment of related concerns will be carried out, which can improve the comprehensive aspect of generating an idea of how literature uses nature to spread consciousness. The impact of eco-feminist approaches is to be deliberated upon, along with a focus on such postmodern eco-feminist texts as *Wide Sargasso Sea.*

II

Understanding Eco-criticism and Eco-feminism

> *Let su permit to nature o have her way*
> *she understands herself better than we do*
>
> – Michael de Montaigne

The words *eco* and *critic* have been derived from the Greek words *oikos* and *kritis*, respectively. When used together they can be translated as 'house judge' which when elaborated refers to *oikos* as nature, and according to Edward Hoagland is the widest home of man and *kritos* as the 'arbiter of taste who wants the house kept in good order, no boots or dishes strewn about to ruin the original décor.'[1] An eco-critic can be defined as a person who can judge various writings based on their faults and merits. These

DOI: 10.4324/9781003383970-8

writings reflect on how culture affects nature with a celebratory view towards it while berating those who spoil it and reverses harm upon it through political activities. Therefore, eco-criticism is a practice that enhances individuals to call out the perpetrators of nature and to incite investigation against the harm done to the natural environment through literary devices.

Ecology as science was connected to the verbal expression of man. However, after World War II and the Great Depression, ecology shifted towards a strong public narrative. It resisted the managerial tools of conservation while crossing boundaries and challenged the support shown to the industries extracting resources as well.[2] The preservationists of ecology utilized this stance to raise the principles of ecology to greater standards to save and safeguard wilderness as well as to show protest against military–industrial research. This leads to the belief that ecology is subversive that serves as a tool of the Leftist political agenda. On the other hand, the eco-feminists started considering science as oppressive and presented the idea that it is an enemy that relies on the male-dominated biological need for sexual reproduction. Eco-feminism is a term that was coined by Françoise d'Eaubonne – a French feminist in the year 1974. She used this term in the essay "The Time for Ecofeminism" where she stated that patriarchy dictates and dominates both nature and women.[3] Furthermore, she urged for an amalgamation of the thoughts as well as activism of feminist and environmentalist ideologies to understand the concept better. Philosopher Karen Warren pointed out that the term *eco-feminism* can take different theoretical positions as it is an umbrella term; however, comprehensive knowledge is necessary to understand the significant connections that are present between nature and women or other subordinates being dominated. If society fails to identify and accept these relations then it leads to an inadequate form of environmental philosophy, environmentalism, and feminism. Eco-feminism is a compelling theory to follow as it has complex notions as to its characteristics for the analysis of the causes leading to environmental issues along with the remedial measures that have been suggested to give it a proper shape.[4] The analysis provided by the eco-feminists spreads awareness based on the need for integrated cultural, social, and economic aspects along with environmentalist thoughts. These differences lie in the national, regional, as well as global contexts and have to be identified for the conservation of the environment. This is how the eco-feminists have shifted from the criticism of anthropocentrism to the conflict between humans and nature. However, further classification and categorization of the environmental analysis need to be done based on the man-made issues of class, ethnicity, gender, and race.

On a more socio-economic note, the eco-feminists have also declared along with the scholars of environmental justice that the entire humankind is not responsible for the damage and destruction that is inflicted upon the environment.[5] The responsibility has to be placed predominantly on those social settings and individuals who hold a position in the power relations of the socio-economic sectors of society and are, therefore, able to make profitable political decisions from the issue. This individual in many societies is largely male elite but not necessarily in all social situations. The relation between humans and nature is not one-dimensional and there are values as well as ethics involved in it too. Christine Cuomo therefore states that:

> This is relevant to its adequacy as an environmental ethic, as the ethics of human interactions with each other determine, are determined by, and are necessarily related to the ethics of human interactions with our environment and its members. Ecofeminists argue that these ethical spheres are inseparable.[6]

Various aspects of eco-feminism have taken considerable shape in various countries around the world such as the United States, Australia, European, Latin American, and North American nations. More strategies have been suggested by feminists for the development of eco-critical notions cross-culturally along with the ethical contexts it offers and the eco-justice related issues. Their focus is to raise awareness in sectors such as Asia, South America, as well as Africa because the knowledge regarding the relationship between history and environment is necessary to promote the goals of eco-feminism and eco-criticism too.[7]

Eco-feminism has taken the shape of a political and social movement as there are various social, political, economic, and cultural connotations related to the concept of environmental degradation as well as environmental conservation. The United Nations and similar international and national organizations have taken up the issue of the degradation of the environment. They have implemented laws and policies to save Mother Nature from the violent grasp of globalization, consumerism, and industrialization. The degradation of nature and the desecration of natural resources have been condemned by world leaders too. However, treaties, conferences, and acts such as the 1992 United Nations Conference on Environment and Development, the Fourth United Nations Conference on Women in Beijing in 1995, the World Summit on Sustainable Development of 2002, the Johannesburg Declaration on Sustainable Development, EU Gender Mainstreaming DGXI Environmental Justice movement have all manifested the idea that the environment needs to be saved.[8] The literary studies must shed light on the improvement of textual knowledge, which can help in the identification of notions of eco-criticism in a distinctive manner. Similarly, the contemporary knowledge of the humans regarding environment and culture has to be materialized as well. Comprehensive knowledge about cultural ecology, therefore, will improve the feasibility of the notion that human culture and different ecological processes as well as the natural energy cycles are not separate but are interdependent and transfused by one another.[9] Along with this, it also recognizes the dynamics that are self-reflexive as well as independent belonging to the cultural processes. Culture is dependent on nature and nature is ineradicably present within the culture of society. This aspect has gained greater attention on an interdisciplinary level which has channelled the appearance of the difference between the natural evolution as well as the cultural evolution that the cultural ecologists accept too. Various literary texts have, therefore, explored and recognized the complex relationship that prevails between the cultural systems and the indicators of the human and non-human aspects of nature.[10] Such grounds have been laid out by the eco-feminist novels that have not only presented various environmental theories but have also devised plots that show the development of humans and nature on a collective front.

III

Literary Criticism and Eco-feminism

The literary criticisms presented by eco-feminists induce within their literary pieces an interdisciplinary display of various approaches. It contributes to these approaches too and shifts from literary criticisms to theories of eco-feminism, feminist philosophies and ethics, critical animal studies, radical environmental philosophy, as well as radical political and economic theories regarding politics, economics, globalization, and ecology.[11] Eco-feminism is, therefore, the intersection of the movements and the analysis pertains to both feminism and ecology. The development of the efforts by activists along with the increase in social consciousness in the early to mid-1970s in North America as well as Europe marked the arrival of eco-feminism in these countries. These efforts focused on the requirement of treating the environment fairly and giving equal rights to women too. The key link between the two concepts as stated above as well lies in the domination of both nature and women by the patriarchal society and especially by the Euro-Western world.[12]

IV

Undertaking and Examining Eco-criticism and Eco-feminism Aspects in Wide Sargasso Sea

While eco-feministic pathways comprise of social movements and activism, global eco-feminism, religion, and academia, a proper reading of the novel *Wide Sargasso Sea* by Dominica-based novelist Jean Rhys is the chosen eco-feministic academic approach in this case. The novel shows a symbiotic relationship that is present between the non-human aspects of the Caribbean land and the humans residing in it. According to Gayatri Spivak, *Wide Sargasso Sea* is a 'cult text' of the English feministic ideologies.[13] It not only questions the perspective of knowledge production based on the Englishness of an individual but simultaneously puts the notion of Englishness into a crisis to be faced by the characters and readers alike.

Reflecting on this thought, the focus needs to be shifted on the concept of identification of self which is deliberated upon throughout the text; 'So between you, I often wonder who I am and where is my country and where do I belong and why was I ever born at all.'[14] The identity crisis faced by Antoinette, the protagonist of the novel who is a Creole woman, becomes unbearable as she is not only detached from her known surroundings to be placed in a strange country and inside the attic of a 'Great House' but also given a different name altogether (Bertha Mason), resulting in her losing all her self-reflective thoughts in the process. However, the one thing that the novel and the protagonist never forget is her nature and animal-consciousness and how nature shapes the life of the people in Jamaica. Antoinette and nature are seen as the same throughout the novel by the outsider Rochester. Therefore, his feelings and thoughts towards both the human and the non-human parts of the Windward Islands are placed accordingly. Antoinette states, 'I thought that when I saw him and spoke to him would be wise as serpents, harmless as doves.'[15] This establishes the fact that not only Rochester considers Antoinette as the Other but she does it too, which is why she claims that she would change eventually after meeting him. Rochester, the English gentleman and Bertha's husband, comments on the strangeness of the place and his wife:

> 'What an extreme green,' was all I could say [...]. Everything is too much, I felt as I rode wearily after her. Too much blue, too much purple, too much green. The flowers too red, the mountains too high, the hills too near. And the woman is a stranger. Her pleading expression annoys me.[16]

Rochester is not only annoyed with the exotic islands but with his wife too who has more similarity to the wildness of the place than an English woman. Rochester is overwhelmed but not awed with his surroundings but irritated, and that irritation is portrayed in his behaviour with his wife. As Antoinette is illustrated as a woman belonging to her natural surroundings, Rochester transfers his feeling of uneasiness about the place towards his wife, which leads to him being indifferent to his wife and finally hating her.[17]

The novelist presents Rochester as the power-hungry and domineering patriarch who tries to dominate both his wife and nature. This imagery follows a similar pattern that was described by C.L.R. James in *The Black Jacobins* regarding the response of the European immigrants towards the lush natural environment of the San Domingo Islands.[18] The same pattern and the same connection between women and nature are detailed in this novel too. Jamaica and the islands that Antoinette resides in are important to her as is the notion of place important for the novel. The place is part of her cultural ecology and she in turn is part of it too. She belongs to Coulibri:

> I am safe. There is a corner of the bedroom door and friendly furniture. There is the tree of life in the garden and the wall green with moss. The barrier of the cliffs and the high mountains. And the barrier of the sea. I am safe. I am safe from strangers.[19]

The institution of marriage and the position of women in it are also questioned in the novel. Antoinette's husband treats her with indifference and engages in multiple relations with other women. They both find each other strange. For Antoinette, only her natural circumstances are true and provide her with motherly warmth. She is closer to nature than her biological mother:

> And though her biological mother is a heartless and cold mother that had to tolerate her for nine months in her bosom, Mother Earth holds her dear and helps her to elevate and blossom like when she says 'If you are buried under a flamboyant tree,' I said, 'your soul is lifted when it flowers. Everyone wants that.'[20]

Through Antoinette, Rhys creates harmony between women and nature and illustrates Rochester as the exploitative Western man who considers himself a superior being based on his race and gender. He is characteristically the detached individual who ultimately annihilates the harmonious relationship.

Further comparison between the relationship of nature with Antoinette and Tia with Rochester's relationship with nature shows that although the Caribbeans are the Other, they still can communicate with nature better.[21] James exemplified how the white immigrants in San Domingo were soon bored of the islands and regarded the women under an evil influence that has materialized from the colonies, slaves, and nature. Rochester initially considers the physical beauty of his wife 'she might have been any pretty English girl' but still refers to her as one with 'sad, dark, alien eyes.'[22] Rochester's attitude towards his wife reflects exactly the environmental ethics of dominating humans as stated by Erika Cudworth who was concerned about how the ethics were related to the 'intrinsic value' of the species and objects of nature. The dismissive attitude of the man towards both nature and women shows the tendency of the Westerner to attach value to these aspects according to the amount of good they can do for them.[23]

While the patriarch aims to exploit both the environment and nature, culture and nature are quite different. According to the man, civilization has to tame nature and the cities are his reality while 'your beautiful island seems to me, quite unreal and like a dream.' The sea is 'unreal and like a dream' for Rochester and the city is unreal and 'like a dream' for Antoinette.[24] However, Antoinette is not an Englishwomen either: she is 'Not quite English and not quite "native"'; Rhys's Creole woman straddles the embattled divide between human and savage, core and periphery, self and Other. The British colonial considers her a part of the Creole culture as Antoinette explains:

> a white cockroach. That's me. That's what they call all of us who were here before their people in Africa sold them to the slave traders. And I've heard English women call us white niggers.[25]

She faces a continuous identity crisis as her land rejects her, as does Tia – a former slave – in whom she sees herself 'Like in a looking glass.' She experiences existential crisis too when she is removed from her motherland and dreams of escaping continuously. Rochester, therefore, hates her as he hates her land. He considered her desire for true love as her promiscuous or evil nature that he connects to the exotic environment of which she is a part. According to the eco-feminists, the patriarch considers nature and women equally, which is the reason for changing Antoinette's name and locking her in the attic.[26] He exploited, marginalized, and oppressed her too; married another woman; and drove her to madness and ultimately suicide.

V

Conclusion

How the women are being marginalized by the patriarchs of the society is similar to how nature is being destroyed. Eco-criticism has the fundamental duty to identify and present how nature is being treated by man through textual references. Eco-feminists have a similar stance but they add the domination of women to the context too as both nature and women are subjected to continuous cruelty. They look forward to a better future with equal rights for women and a sustainable developmental programme for nature. Texts such as *Wide Sargasso Sea* are valuable as they bring forth the threat that women and nature perceive. Identification of the underlying issues and eradicating them from society on an urgent basis is what the eco-critics and the eco-feminists urge their readers to do. The key concern is not to detach nature and culture from humanity but to promote a harmonious relationship between the three concepts.

> 'But how can rivers and mountains and the sea be unreal?' 'And how can millions of people, their houses, and their streets be unreal?' 'More easily,' she said, 'much more easily. Yes, a big city must be like a dream.' 'No, this is unreal and like a dream.'
>
> (*Wide Sargasso Sea*, 73)

Notes

1 William Howarth, "Some Principles of Ecocriticism." *The Ecocriticism Reader: Landmarks in Literary Ecology* (1996): 69–91.

2 Howarth, "Some Principles of Ecocriticism." 69–91.
3 Sylvia Mayer, "Literary Studies, Ecofeminism and Environmentalist Knowledge Production in the Humanities." *Nature in Literary and Cultural Studies: Transatlantic Conversations on Ecocriticism* 3 (2006): 111–127.
4 Mayer, "Literary studies, ecofeminism and environmentalist knowledge production in the humanities." 111–127.
5 Mayer, 111–127.
6 Mayer, 111–127.
7 Greta Gaard, "New Directions for Ecofeminism: Toward a More Feminist Ecocriticism." *ISLE: Interdisciplinary Studies in Literature and Environment* 17, no. 4 (2010): 643–665.
8 Susan Buckingham, "Ecofeminism in the 21st Century." *The Geographical Journal* 170, no. 2 (2018): 146–154.
9 Hubert Zapf, "Ecocriticism, Cultural Ecology, and Literary Studies." (2010): 136–147.
10 Zapf, "Ecocriticism, Cultural Ecology, and Literary Studies." 136–147.
11 Greta Gaard, "Strategies for a Cross-Cultural Ecofeminist Literary Criticism." (2010): 47–52.
12 Jane Eyre, "An Ecofeminist Approach to Jean Rhys's Wide Sargasso Sea." *MP: An Online Feminist Journal* (2008): 8–18.
13 Laura E. Ciolkowski, "Navigating the Wide Sargasso Sea: Colonial History, English Fiction, and British Empire." *Twentieth Century Literature* 43, no. 3 (1997): 339–359, doi: 10.2307/441916.
14 Ibid.
15 Jean Rhys, "Wide Sargasso Sea. Nueva York." (1966): 161.
16 Eyre, "An Ecofeminist Approach to Jean Rhys's Wide Sargasso Sea." 8–18.
17 Ibid.
18 Ibid.
19 Zahra Sadat Ismailinejad, "Jean Rhys's Wide Sargasso Sea: An Ecocritical Reading." *International Letters of Social and Humanistic Sciences* 49 (2015): 146–154.
20 Ismailinejad, "Jean Rhys's Wide Sargasso Sea: An Ecocritical Reading." 146–154.
21 Ismailinejad, 146–154.
22 Eyre, 8–18.
23 Eyre, 8–18.
24 Ismailinejad, 146–154.
25 Ciolkowski, 339–359.
26 Ismailinejad, 146–154.

Works Cited

Buckingham, Susan. "Ecofeminism in the 21st Century." *The Geographical Journal*. 170.2 (2018): 146–154.

Ciolkowski, Laura E. "Navigating The Wide Sargasso Sea: Colonial History, English Fiction, And British Empire." *Twentieth Century Literature*. 43.3 (1997): 339. doi:10.2307/441916.

Eyre, Jane. "An Ecofeminist Approach to Jean Rhys's Wide Sargasso Sea." *MP: An Online Feminist Journal*. (2008): 8–18. http://academinist.org/

Gaard, Greta. "New Directions for Ecofeminism: Toward a More Feminist Ecocriticism." *ISLE: Interdisciplinary Studies in Literature and Environment*. 17.4. (2010): 643–665.

Gaard, Greta. "Strategies for a Cross-Cultural Ecofeminist Literary Criticism." (2010): 47–52. OUP.
Howarth, William. "Some Principles of Ecocriticism." *The Ecocriticism Reader: Landmarks in Literary Ecology*. University of Georgia. (1996): 69–91.
Ismailinejad, Zahra Sadat. "Jean Rhys's Wide Sargasso Sea: An Ecocritical Reading." *International Letters of Social and Humanistic Sciences*. 49 (2015): 146–154.s.
Mayer, Sylvia. "Literary Studies, Ecofeminism and Environmentalist Knowledge Production in the Humanities." *Nature in Literary and Cultural Studies: Transatlantic Conversations on Ecocriticism*. 3 (2006): 111–127.
Rhys, Jean. *Wide Sargasso Sea*. New York: W.W. Norton. (1966).
Zapf, Hubert. "Ecocriticism, Cultural Ecology, and Literary Studies." (2010): 136–147. https://ecozona.eu. Wahington.

7 Studying 'Cli-fi'

Thinking about the 'Unthinkable'

Sonam Jalan

Climate change has become a harsh reality of our times from which we cannot escape. It has become a part of our everyday lives and as such adaptability is the only way to mitigate the impact of the changing climate. Since the late 1980s, global warming has generally been recognized as one of the greatest challenges facing humanity in the twenty-first century. Climate change induced by global warming is affecting the earth's ecosystem thereby threatening the future of our planet, resulting in substantial species extinction, increased food insecurity, frequent warming and acidification of oceans, sea-level rise, frequent and long-lasting heatwaves, and so on. Staggering under the adverse effects of climate change requires a change in our collective consciousness as only then can we change our behaviour towards this rapidly approaching crisis. With this recognition, a huge corpus of writing on climate change has flourished in the past few years. The emergence of 'Cli-fi,' an abbreviation in analogy with sci-fi, apparently coined by the journalist Dan Bloom in 2007, as a new genre of fiction with a 'moral imperative' is promoting engagement with a global issue. This has become the subject of numerous blogs and reading forums on the internet and has brought into focus a growing academic interest.

What is 'Cli-fi'? Though there is no proper definition, 'Cli-fi' provides a convenient term for a significant body of narrative work, the so-called 'serious fiction' that would deal with a shift from a close focus on domestic realism to a global view of environmental vulnerability. Originating from the broad area of science fiction, these novels are written with a moral sense of what might be the consequences if we do not stop the human-induced climate change and global warming. These narratives explore the anthropogenic climate change not just in terms of setting but explore psychological, social, and economic issues, combining fictional plots with scientific facts, speculations on the future, and reflecting on the human–nature relationship. Lawrence Buell defines climate narratives as a work in which 'The nonhuman environment is present not merely as a framing device but as a presence that begins to suggest that human history is implicated in natural history' (Buell, 1995, p. 7).

DOI: 10.4324/9781003383970-9

So, can these books save the planet? Are these climate narratives capable of moving the people aesthetically by making them imagine the impacts of climate change? In my opinion, yes, they are. These climate narratives are capable of changing the mindset of the readers by engaging them in a way that the statistics of scientists cannot. Barbara Kingsolver's *Flight Behaviour* creates an emotional resonance with the reader through a fictional framework warning people of the adverse effects of the warming planet on the lifestyle of the Monarch Butterflies setting his characters amidst familiar family tensions. The most challenging part of climate change is that the apocalyptic consequences of our carbon use are emerging gradually and globally, rather than in a single newsworthy disaster. Set in the future, in a world where advanced climate change has wreaked irreversible damage upon the planet, these novels portray terrible futures, drowned cities, uncontainable diseases, burning worlds – all such scenarios that scientists have long tried to warn us about. These serve as a warning to the readers by moving them aesthetically.

By making these abstract futures more immediate, fiction explores the possibility of fighting and adapting to the changing climate. Despite being fictitious, these narratives inculcate two very important components – facts and fear. The facts are rooted in scientific evidence that the human-induced climate change is not only real but increasing rapidly in both quantity and consequence. The fear of human beings suffering in the aftermath of an eco-apocalypse or eco-catastrophe is not only real but morbidly relatable. Based on these facts writers are strengthening the awareness regarding the crisis faced by our planet and compelling them to find out ways of adaptation. In some narratives, the protagonists discuss climate change and, in a few cases, they are scientists trying to come up with solutions. Thus, Frederick Buell comments on the novels of the Anthropocene as:

> Environmental Crisis appears in recent literature not just as a foregrounded theme but much more complexly and actively as part of writers' construction of their characters' psyches, thoughts and actions; writers' creation of fictional conflicts and plots; and writers' crafting of narrative structure, voice and other aspects of style.
>
> (Buell, 261)

My argument in this chapter is that the fictions on climate change move beyond the stereotypical notions of ecocriticism and offer a rethinking and reimagining of the contemporary environmental problems. Ecocriticism has been defined by Cheryll Glotfelty as the relationship between literature and the physical environment that has an earth-centred approach to literary studies. Emerging in the 1980s on the shoulders of the 1960 environmental movement that began with the publication of Rachel Carson's *Silent Spring*, ecocriticism has been and continues to be an 'increasingly heterogenous movement' (Buell, 1995). Since the 1990s ecocriticism has diversified into specializations like affective and empirical studies taking into account climate change as well as the current coronavirus pandemic. Nature has always been an issue of concern for the eco-critics. There were nature-related novels earlier by Hardy, Austen, and Lawrence, but in post-1975 dystopian works (the year when the term 'global warming' was coined by Wallace Smith) we see how the techno-utopian solutions ultimately fail to address the root causes of the climate crisis, and hence a solution to the problem can only be achieved through a social change.

Eco-fiction is fiction where nature and the relationship between humanity and their physical environment play a major role in the plot. It acts as an umbrella term that includes several other terms like 'Cli-fi,' ecopunk/solar-punk, speculative fiction, and Anthropocene fictions. These narratives offer an opening to address wider environmental issues. 'Cli-fi'/Climate fiction focuses specifically on the man-made climate change and the anthropogenic impacts instead of the unstoppable natural disasters such as volcanic eruptions or the meteorites hitting the earth.

Before proceeding further, let us analyze the germination of climate fiction and its history. In all probability, it began with J.G. Ballard's *The Drowned World* (1962), a novel about global climate change resulting from human exploitation. However, we can go further back to Jules Vernes's *The Purchase of the North Pole* (1889) from where sci-fi began. We can even go further back to Mary Shelley's *Frankenstein* where we witness several climatic changes in the background. Another work that needs a mention is John Steinbeck's *The Grapes of Wrath* (1939) which depicts the condition of climate migrants through its story of a poor family of tenants driven out of the 'Dust Bowl' – another large-scale environmental catastrophe. It has been predicted that climate change could create as many as 143 million 'climate migrants' by 2050.

The roots and iconography of the climate novels belong to science fiction with J.G. Ballard's series of cataclysm: *The Wind from Nowhere* (1961), *The Drowned World* (1962), *The Burning World* (1964), *The Drought* (1965), *Hello America* (1981). These novels describe the environmental catastrophes in pseudo-scientific ways. The story of *The Drowned World* is set in twenty-second century London where the protagonist Dr Robert Kerans, member of a scientific exploration, arrives to carry out his research in the drowned city of London. Fluctuations in solar radiation have resulted into melting of the ice caps, sending the planet into a new Triassic Age of unbearable heat. It is noteworthy that Ballard continues his apocalyptic cycle in a consistent style. Beginning with the devastating hurricane in *The Wind from Nowhere*, to the immense floods in *The Drowned World*, he moves on to fatal aridity in *The Drought*. It is exceptionally striking that in all of these apocalyptic novels, Ballard places the reaction of human characters at the core concentrating on the devastating impacts of the disasters portrayed.

With this as a starting point we can identify several novels, most notably, science fiction such as George Turner's *The Sea and the Summer* (1987), Kim Stanley Robinson's "Science in the Capital" trilogy (2004, 2005,

2007), and Paolo Bacigalupi's *The Windup Girl* (2011). The emerging literary climate change fictions starting with Maggie Gee's *The Ice People* (1998) and *The Flood* (2004), T.C. Boyle's *A Friend of the Earth* (2000) gained momentum with Margaret Atwood's 'Cli-fi' trilogy (*Oryx and Crake*, *The Year of the Flood*, and *Maddaddam*), often recommended by many 'Cli-fi' fans as a great place to start when exploring the genre. The trilogy takes place in the near-future where plagues, floods, and genetic engineering have completely altered the world that we know today. To name a few more there are Jeanette Winterson's *The Stone Gods* (2007), Cormac McCarthy's *The Road* (2006), Ian McEwan's *Solar* (2010), James Miller's *Sunshine State* (2010), Nathaniel Rich's *Odds against Tomorrow* (2013).

Environmental apocalypse is not a brand-new literary concern since we have already witnessed the impassive earth as a motivating image in many works. We can trace humanity's relationship with the environment all the way back to the early stages of creation with the formation of the Holy Books. The first book of the Bible asserts God's dictate on man to master and subjugate all living forms on the surface of earth. However, the scenario was completely different in non-Western tribal cultures with their close affiliation towards nature. The ideas concerning nature and their representation dates back in the works of Geoffrey Chaucer in the medieval period whose understanding of nature was not limited to its physical locations alone rather imbibed its religious and spiritual notions also. During the Elizabethan period, extensive use of nature imagery is seen in the works of William Shakespeare and his contemporaries. However, the actual forerunner of early ecocriticism can be traced back in British Romanticism with the poems of William Wordsworth, Coleridge, and Keats, reflecting humanity's strong and indispensable tie with nature. With the emergence of Industrial Revolution in Britain along with remarkable scientific and technological breakthroughs, the rustic environment of the Romantic period was taken over by an urban, industrial, and a completely different environment.

Charles Dickens, one of the leading authors of that time, was seen emphasizing on the effects of industrialization on environment. His novels such as *Bleak House*, *Little Dorrit*, and *Our Mutual Friend* all share a shocking description of ecological disasters in sharp contrast to the glorification of nature in the works of his predecessors. His writings proved that fiction is capable of reflecting the hardships of life in an ecologically ruined society, thereby creating awareness among its readers for the preservation of nature and re-establishment of climate justice. With Rachel Carson's *Silent Spring*, the content of ecology witnessed a transformation from passive reflection to active intervention. Antonia Mehnert in her book *Climate Change Fictions: Representations of Global Warming in American Literature* (2016) investigates the portrayals of climate change in literary fiction. She defines climate change fiction as

> literature dealing explicitly with anthropogenic climate change – gives insight into the ethical and social ramifications of this unparalleled environmental crisis; reflects on current political conditions that impede action on climate change, explores how risk materializes and affects society, and finally plays an active part in shaping our conception of climate change.
> (Mehnert, 4)

Such fiction forces environmental criticism to move beyond its long-standing interest in concepts of 'nature' and 'place' and embrace a new understanding of the local in relation to the global. Considering such complexity and uncertainty at the source of climate change fiction, this book moves beyond a solely ecocritical analysis and shows how such fiction constitutes an insightful cultural information for discussion in the environmental humanities in general.

The extraordinary changes in the weather patterns induced by human emission of greenhouse gases have caused immense environmental and social changes. These effects of climate change are interrelated with each other, increasing their violence, putting at risk the species that inhabit the earth including humans. Variations in the physical systems of the planet, including the melting of the poles causing glacial regression, snow melting, warming and thawing of permafrost, flooding in rivers and lakes, droughts, coastal erosion, the rise of sea levels, have destroyed the economic means of subsistence. Climate change destroys food crops and freshwater supply. Juxtaposed with extreme weather events, it leads to catastrophic effects on human health, destruction, and loss of economic livelihood causing migration of many climate refugees.

Climate refugees are the millions of people who have been displaced by the environmental disasters brought on by climate change. Legally, a refugee is a person who has been forced to flee his or her country because of violence, persecution, or war. However, the growing awareness of the impact of climate change has proved that it has as much potential to displace communities as violent conflicts. A 'climate migrant' or 'ecological refugee' is someone who is moving away from the coast because his home has been wiped out due to the enraging floods; he is a farmer in a country where it is too hot and too dry and unable to subsist anymore migrates to a city. These movements are not just within the country but across international borders as well. This is not at all a pretty picture. According to the UN Refugee Agency, over 25 million people are forced to leave their homes due to disasters every year.

Climate fictions provide a realistic picture of the tales of a planet in turmoil which help us imagine the plight of people whom flood or drought displaces. This indirectly raises radical empathy. Literature acts as a catalyst for climate action by bringing science to the masses, translating abstract charts and temperature variations into engaging human experience. It triggers and enriches debate and encourages people to reflect on what is important for them. Adam Trexler's *Anthropocene Fictions: The Novel in a Time of Climate Change* (2015) makes a significant contribution to the field of climate fiction. Here Trexler identifies the evolution of the climate novel by considering novel as 'a privileged form to explore what it means to live in the Anthropocene moment.'

It is to be noted that climate change differs from other environmental problems on the amount of risk involved. Climate change that takes place due to increase in carbon dioxide concentration is largely irreversible for 1,000 years even after the emission stops. Ulrich Beck talks about a global risk society that is caused due to constant technological change which requires us to respond and adjust to these changes. This risk, he argues, includes a whole range of interrelated changes within contemporary social life such as shifting employment patterns, heightened job insecurity, declining influence of tradition and custom erosion of traditional family patterns and democratization of personal relations. The ecological crisis is central to this analysis of the contemporary period. He asserts that humans have always been subjected to a level of risk such as the natural disasters – caused by the non-human forces. Modern societies are however subjected to risks such as pollution, newly discovered illness, crime – resulting from the process of modernization itself (Beck, 1992). Anthony Giddens describes such kind of risks as manufactured risks. Since these risks are the product of human activity, authors like Giddens and Beck argue that it is possible for societies to assess the level of risk that is being produced or that is about to be produced.

Amitav Ghosh's latest work *Gun Island* is set in an ecologically unstable world that explores several environmental crises like wildfires, hailstorm, extinction, and migration. Deen, a New York-based antiquarian book dealer arrives at the Sundarbans, the (disappearing) wetlands wedged between India and Bangladesh, in search of a shrine. He is also investigating the truth behind the myth of the Gun Merchant and Manasa Devi, the goddess of snakes. Ghosh weaves the myth of the Gun Merchant into contemporary weather-related realities such as the Los Angeles wildfires, the unusual travels of dolphins and spiders, and the sinking buildings of Venice, to create an ultimately hopeful tale of our times. While researching for his earlier novel *The Hungry Tide* in the early 2000s, about the ecosystem of the Sundarbans that support the endangered Bengal tiger and thousands of other species, Ghosh could already see the impact of climate change with bigger waves and worsening cyclones hindering farming and the crisis of survival of the local communities. This shift, over the years, has directly or indirectly forced a major cross-section of the 4 million inhabitants of the Sundarbans to flee the land. Scientists assume that the low-lying regions like the Lakshadweep and the Sundarbans will be worse hit, seriously affecting the lives of millions of people especially the local communities.

The global increase in temperatures is not only influencing the physical and human systems of the planet but is also causing a severe impact on its biodiversity. There is the death of flora and fauna in terrestrial and marine ecosystems, wildfires, and change in migratory patterns. Barbara Kingsolver's *Flight Behaviour* is set in the rural countryside where a massive butterfly die-off brings in scientists from the big cities to study the problem. These butterflies had migrated to Feathertown due to global warming. Along with such serious fiction, there is Ian McEwan's *Solar*, a comic tale to portray a sudden turn in the protagonist's life from being a greedy, selfish urban consumer to an accidental expert on anthropogenic climate change. Here McEwan employs humour and ridicule to exhibit the weaknesses and greed of mankind in a satirical way.

It is interesting to note that while the South Asian climate narratives are deeply connected with other social and political struggles sustaining a local, relevant, and more nuanced focus on climate change, in the Western context, climate novels remain largely conventional with crumbling cities and dystopian systems of power. These books depict an isolated protagonist attempting to survive in these wastelands. There is a shift in focus from the environment towards struggle for survival.

As a genre 'Cli-fi' has certain limitations due to its reliance on apocalyptic scenarios and its didactic tendency. Cultural narratives of climate change are often disregarded in public discussions as climate change is considered an issue of concern solely for natural sciences and engineering practices. This issue has been epitomized in C.P. Snow's *The Two Cultures* (1959) which talks about the historically entrenched division of nature and culture in Western societies which determines who possesses the legitimate power to address issues related to nature. For decades, science has shown that humans will suffer dire consequences if we do not implement sustainable practices and decrease our overall carbon footprint. The question is, in what ways is the humanist approach towards the problem different from their scientific counterparts?

While the scientists and engineers utilize their backgrounds and respective fields to develop sustainable technologies, literary humanities tend to write policies, produce news, and educate the masses. Their writing craft makes it easy to relate scientific data to the general public and inspire changes through persuasive pieces. Hence, there is a need for collaboration between the two cultures. The 2018 IPCC (Inter-Governmental Panel on Climate Change) special report on 'Global Warming of 1.5°C' is a recent example of such collaboration. This report was penned down by 91 authors with extensive backgrounds from a variety of fields, including science, philosophy, languages, economics, mathematics, and communication from universities all around the globe.

Amitav Ghosh laments that it is the literary world's greatest impediment in being unable to capture the realities of climate change in a novel. Being equally serious and analogous to war, slavery, colonization, famine, and other crises and events that have seeped into various forms of literature through ages, Ghosh believes that manifold ways should be devised to address such a cataclysmic occurrence. His non-fiction, *The Great*

Derangement: Climate Change and the Unthinkable, questions the absence of such an urgent phenomenon in fictional writing. He meditates upon the 'uncanniness' of the apparently 'improbable' climate change which makes it difficult for a writer to use it in a fictional framework contributing to humanity's derangement. He admits that whenever improbable episodes appear in literature they are segregated as surreal or magic real.

The greatest challenge faced by humans today is that they are unaware of the causes and consequences of climate change and how to moderate those causes. Storytelling is an important tool that shall help us to adapt to new realities. It gives climate change a human focus by translating complex and evolving scientific concepts into tales. In the wake of the growing concern for the fast-deepening environmental crisis, this chapter thus highlights how climate fiction can effectively be a powerful mediating tool in understanding the dramatic and emotional contours of climate change. With the effects of climate change looming, it only makes sense that we will need more ways to help us understand how our world is changing.

Works Cited

Beck, U. *Risk Society: Towards a New Modernity*. London: Sage, 1992.

Buell, L. *The Environmental Imagination: Thoreau, Nature Writing and the Formation of American Culture*. London, England: Harvard University Press, 1995.

Wells, D.W. *The Uninhabitable Earth: A Story of the Future*. Penguin Books, 2019.

Part II

Ecology and Literary Representation

8 Gerard Manley Hopkins – A Priest of Ecology

Goutam Buddha Sural

F.R. Leavis in his *New Bearings in English Poetry* (1932) claimed that '[Hopkins] is likely to prove for our time and future, the only influential poet of the Victorian age, and he seems to me the greatest' (156). Whether Hopkins can be acknowledged as the 'greatest' poet of the Victorian period may always be contested, but the fact remains that in the Victorian period in the field of green issues Hopkins is possibly the most important poet. Jerome Bump in his article "Hopkins, the Humanities, and the Environment" published in 1974 in *Georgia Review* points out that poetry can play a useful role in interdisciplinary environmental science. According to him, poets in many ways have anticipated environmental scientists and have expressed their ideas in definite, convincing, and forceful language (230). But discussions on the mutual relationship between poetry of the nineteenth century and ecology are not many and if one goes back to Victorian literature such discussions are even less. Bump points out that 'Some writers on environmental issues mention Dickens's *Hard Times*, of course, and sometimes Ruskin, but I have yet to find anybody who is even aware of the contributions of, say, Gerard Manley Hopkins' (229). He further wrote, 'Hopkins' affinities with the modern ecology movement have hardly been noticed in the more than fifteen hundred books and articles on him which have appeared since the publication of his poems in 1918' (232).

Jonathan Bates's edited book *Romantic Ecology: Wordsworth and the Environmental Tradition* published in 1991 is often regarded as a pioneering work in the field. But the book, as the title suggests, does not focus on Victorian poets, though there are passing references of major Victorian poets, including Hopkins. Bate, however, discusses Ruskin and Morris at length to point out that the core thought behind the ecologies of Wordsworth, Ruskin, and Morris is 'modern' (60–161). In the twenty-first century, there is, however, a renewed interest in Hopkins's poetry and several articles appeared in different international journals focusing on the ecological aspects in Hopkins's poetry. In 2002 came out a book titled, *The Environmental Tradition in English Literature*, edited by John Parham in which the editor himself contributed an essay, "Was There a Victorian Ecology?" In that essay, Parham discusses many Victorian poets and

DOI: 10.4324/9781003383970-11

finally asserts that 'Hopkins was the one Victorian poet who consistently, imaginatively re-created the specific conditions of the Victorian ecosystem' (170–171). He strongly argues that Hopkins's poems 'contain an imaginative seed with the essence, not just of a broader understanding of Hopkins, or of Victorian ecology, or ecocriticism, but of ecology itself' (171).

In another article, "Hopkins' Spiritual Ecology in 'Binsey Poplars,'" published in *Victorian Poetry* in 2004, Brian J. Day observes that even though 'Hopkins's poetry offers us insights into our own twenty-first century ecological discourse' and his 'ecological sensibility flows into the broad stream of ecological nature writing … his place in the canon of ecological thinkers has been largely ignored' (190). According to Day, Hopkins's ecology is 'fundamentally spiritual' (181). The tripartite relationship among nature, the created world, and God is central to Hopkins's poetry. The most interesting publication is a book, published in 2010, by John Parham whom I have already mentioned. The book's title is interesting – Parham substitutes the first two letters of Hopkins's first and middle name and the title of the book is *Green Man Hopkins: Poetry and the Victorian Ecological Imagination.*

There are a number of poems by Hopkins which exemplify a perfect agreement between Christian religiosity and his sense of deep ecology – "God's Grandeur," "Spring," "The Sea and the Skylark," "Pied Beauty," "Binsey Poplars," "Inversnaid," "Ribblesdale." In many of these poems, Hopkins celebrates his love for nature and his sensuous pleasures. But at the same time, his poems reflect his distress and anxiety at the destruction of nature and landscape. In poems like "God's Grandeur" and "Binsey Poplars," Hopkins unambiguously writes about the pervasive impact of the Industrial Revolution on the environment. The two poems clearly voice a protest against soil pollution and deforestation and critique the destruction of nature by man in the name of industrialization. We have to remember here that Hopkins was neither averse to modernization nor was he against industrialization *per se* but he was against the mindless expansion of industry and urbanization. Jude Nixon explains that:

> Hopkins's conversion to Roman Catholicism (1866) and Jesuit affiliation (1868) granted him membership in a religious community open to a scientific inquiry on the ground that science reveals the mystery of the universe and is ultimately compatible with Church dogma. Issues raised by the biological and geological sciences unsettled Victorians as well as Roman Catholic Victorians, like St. George Mivart, who imagined that there were in fact solutions within emerging scientific theories not incompatible to faith.
>
> (132)

This combination of faith and science is always found throughout Hopkins's poetry. "God's Grandeur" begins thus:

> The world is charged with the grandeur of God
> It will flame out like shining from shook foil;
> It gathers to a greatness, like the ooze of oil
> Crushed.
>
> (ll, 1–4)

The world is the world of nature, the world of created beauty which is pervaded by the power of God, but the power is latent and it may reveal itself any moment in the manner of the sudden flash of lightning. Hopkins compares the momentary flash of God's power with the sudden flash of a gold foil shaken in the sunlight. The opening lines of the poem closely echo Hopkins's commentary on the *Spiritual Exercises* of St. Ignatius Loyola:

> The last mystery meditated on in the *Spiritual Exercises* is our Lord's ascension ... it is the contemplation of the Holy Ghost sent to us through creatures ... All things, therefore, are charged with love, are charged with God, and if we know how to touch them give off sparks and take fire, yield drops and flow, ring and tell of him.
>
> (*Sermons*, 129)

The above account is almost a prose rendering of the opening lines of the sonnet. The same light imagery is at work. In his essay "Hopkins and the Religion of the Diamond Body," published in *The Cambridge Quarterly* in 1998, Keith Sagar observes:

> The fourth word is charged with meanings, and plunges us deep into Hopkins' distinctive poetic and religious world. The primary meaning is, of course, electrical. It suggests that God flows through the world as an energizing current, that everything in creation is therefore connected to everything else, part of the same circuit, and also to God, the source. It also suggests that the world is a huge battery in which creative energy lies latent, in which is stored an infinite potential for renewal.
>
> (18)

But the poem makes the reader feel the negative impacts of industrialization. Man can recognize God's power only if he is aware that the world of created beauty is a manifestation of God. Unfortunately, it is this awareness which people lack, and the poet grieves: 'Why do menthen now not reck his rod?' and in the next four lines the poet laments of what man has made of man and the resultant mangling of nature by man. The repetition of the phrase 'have trod, have trod, have trod' when read aloud conveys to the reader the sense of havoc caused by industry. As John Parham observes:

> In the brutal repetition ... of "have trod, have trod, have trod" Hopkins succeeds in conveying industrial impacts rhythmically, not just "on the

> ground" but *in* the ground, the underlying ecosystem, more than just an abstract environment, "disfigure[d]" or altered by a human dominance threatening to create a monoculture ("wears man's smudge and shares man's smell").
>
> (205)

The poem categorically points to the desensitization of human beings caused by industrialization: '... nor can foot feel being sod.' But the negative impact is countered by the poet's assertion that despite this 'nature is never spent' as there 'lives the dearest freshness deep down things.' And the poem moves on to trace the source of the 'freshness' in the Holy Ghost who constantly rejuvenates nature. There is a hidden message which is that if the ecosystem is harmed, humanity will suffer its consequence but nature will be continually refreshed and energized regardless of human society and its industrial practices. The unmistakable note of anxiety expressed by the poet due to the destruction of nature is assuaged by his faith in God, the ultimate source of energy.

Hopkins's poem "The Windhover," about which the poet himself said that it was the 'best thing he ever wrote,' speaks of the bird's energy that has to be 'caught.' The choice of the word caught is significant. The emphasis is not so much on the bird's flight as a spectacle but its power over the wind. Hopkins could have used a verb like 'saw' instead of 'caught,' had he prioritized the visual aspect of the bird's flight. The alliteration in the second line corroborates this view with the repetitive use of the plosive sound 'd.' As the bird is coming out of its nest drawn by the multi-coloured dawn, it unleashes its energy with great gusto and rides over the wind with complete mastery. In this connection, let me quote the words of John Parham. He writes:

> it is, specifically, an ecological sense of the bird's energy that is 'caught' in these first six lines as Hopkins depicts the falcon working with and against the wind. The octave represents, therefore, in the fullest possible sense, an example of an ecopoetry that unifies eco- and technocentric perspectives.
>
> (246)

Claude Colleer Abbott, in his "Introduction" to *The Letters of Gerard Manley Hopkins to Robert Bridges*, argues that in many of Hopkins's nature poems the impressionism of the octave in effect overpowers the exemplum of the sestet:

> They are poems written to the glory of God by a man who is looking on the world as charged with His grandeur and revealing His bounty and presence. But always as I read them I feel that the poet is primarily seized by the beauty of the earth and that though a man of exquisitely

> tempered and religious mind, his senses, not his religion, are in the ascendant … his visions of earth and her creatures make a bevy of astonishing and new felicities rarely to be matched in English poetry [in which the] fusion of earthly beauty and exemplum is often so incomplete that the second is merely the addendum of a poet captive in the first place to the beauty besieging his senses. This loveliness is here for its own sake.
>
> (xxvii–xxviii)

These words of Abbott immediately remind us of Hopkins's poem "Pied Beauty," which may be regarded as a poem celebrating biodiversity. The poem begins and ends with a thanksgiving to God for the dappleness one finds in nature. The poem proceeds through a depiction of a variety of beautiful things which are variegated or contain contrariety of various kinds – colour, taste, speed, brightness – to an affirmation of the Creator of them in whom lies the supreme gift of resolving the paradoxes within his unity.

Hopkins's poem "Binsey Poplars" may be regarded as an 'elegiac, preservationist protest' against the felling of poplar trees at Godstow, Oxford. Trees and flowers were central to Hopkins's philosophy of nature and many of his letters and journal entries bear proof of this. For him, hacking a tree is equal to wiping out the inscape of the world. In one of his journal entries, he records his deep anguish at the felling of an ash tree:

> The ash tree growing in the corner of the garden was felled. It was lopped first: I heard the sound and looking out and seeing it maimed there came at that moment a great pang and I wished to die and not see the inscapes of the world destroyed anymore.
>
> (*Journals and Papers*, 8/4/73: 230)

In another of his letters written to his mother on 19 October 1863, Hopkins expressed his agony in more uncompromising words:

> They have cut down the beautiful beech in the Garden Quad, which stood in the angle of Fisher's buildings because it was said to darken their rooms. This is a wicked thing; such a beech no doubt has not it's like in Oxford, beech being a rare tree here. Its destruction is owing to the Fellows Green[1] and Newman[2]. The former is of a rather offensive style of infidelity, and naturally dislikes the beauties of nature. It is said the Fisher building[s] are to be pulled down […] I wish they could have pulled them down first, and let the tree stand.
>
> (*Further Letters*, 19/10/63: 83)

When Hopkins returned to Oxford as a priest, he lamented in a letter to R.W. Dixon the absence of natural beauty he had known as an undergraduate. He wrote that it was 'already abridged and soured and perhaps will be put out altogether' and, in a postscript to the letter, added 'I have been up to Godstow this afternoon. I am sorry to say that the aspens that lined the river are everyone felled' (*Correspondence: RWD*, 27/2/89: 20, 26). "Binsey Poplars" was composed on the same day.

> *Binsey Poplars*
> felled 1879
> My aspens dear, whose airy cages quelled,
> Quelled or quenched in leaves the leaping sun,
> All felled, felled, are all felled;
> Not spared, not one
> That dandled a sandalled
> Shadow that swam or sank
> On meadow & river & wind-wandering weed-winding bank.
> O if we but knew what we do
> When we delve or hew—
> Hack and rack the growing green!
> After-comers cannot guess the beauty been.
> Ten or twelve, only ten or twelve
> Strokes of havoc únselve
> The sweet especial scene,
> Rural scene, a rural scene,
> Sweet especial rural scene.

Brian J. Day in his article "Hopkins' Spiritual Ecology in 'Binsey Poplars'" observes:

> The tree's suffering is Christ's suffering - in the loss of its Christ-like perfection. This being so, Hopkins is under no obligation to move away from the tree in equating its fate with Christ's. The aspen is "dear" because Christ is dear. The perception of one makes possible the perception of the other. Inscape, therefore, does not move Hopkins away from the thing into the realm of transcendental signifiers but keeps him looking at the thing itself even as he perceives Christ's sacrifice.
>
> (187–188)

The dedicatory note 'felled 1879' conveys the deep sense of loss due to 'the strokes of havoc that unselved the sweet special rural scene.' The first two lines of the poem refer to the process of photosynthesis of the lopped trees. Parham in his book says, '"my aspens dear" immediately gives way to an embodiment of photosynthesis, fundamental to the ecological process.' Day also convincingly argues that the trees can be interpreted as vegetal 'lungs' whose leaves "quell" or "quench" the sun by transforming one form of energy into another through photosynthesis, a concept known by that time and one which Hopkins, 'a contributor to scientific journals and reader of Ruskin ... would have known and understood' (188).

The phrase 'airy cages' also refers to the foliage which is metaphorically compared to cages that imprison the sun rays to maintain balance in the ecosystem. The repetition of the word 'felled' emphatically communicates the destruction caused by humankind to nature and has an obvious reference to Man's 'fall' the sense of which is aptly conveyed by the lines 'O if we but knew what we do/When we delve or hew—/ Hack and rack the growing green!' The words remind us of Christ's words on the Cross, 'Father, forgive them; for they know not what they do' (Luke, 23–34). The unselving of the poplars not only robs the nature of its 'sweet special rural scene' it also suggests that if this special scene cannot be seen by the posterity ('aftercomers') then it is also a sin committed against God as the destruction of the poplars signifies the destruction of the inscape of God. The inscapes that Hopkins found in nature were a means to apprehend the glory of God. For example, on 18 May 1870, Hopkins records in his *Journal*:

> One day when the bluebells were in bloom I wrote the following. I do not think I have ever seen anything more beautiful than the bluebell I have been looking at. I know the beauty of our Lord by it.
>
> (199)

The condemnation of nineteenth-century urbanization and industrialization is explicitly conveyed in poems like "Dun Scotus's Oxford" and "The Sea and the Skylark." While a sense of nostalgia for the cheer and charm of earth's past prime can be felt in these poems they simultaneously disapprove of the destruction of nature for the spread of shabby townships. The poet in "Duns Scotus's Oxford" wrote:

> Thou has a bare and brickish skirt there, sours
> That neighbour nature thy grey beauty is grounded
> Best in; graceless growth, thou has confounded
> Rural Keeping—folk, flocks and flowers.
>
> (ll, 5–8)

The beautiful medieval Oxford surrounded by natural beauty is corrupted by the growth of modern industry and setting up of townships. These have turned Oxford into a 'graceless' town full of ugly structures of bricks. The city now stands de-folked, de-flocked, de-flowered. The lines remind us of Ruskin's lament at nature being spoilt by industrial blight. I may draw your attention here to the words of John Ruskin in one of his lectures, "Modern Manufacture and Design," which he delivered at Bradford in March 1859. Ruskin wrote:

> The changes in the state of this country are now so rapid, that it would be wholly absurd to endeavour to lay down laws of art education for

> it under its present aspect and circumstances; and therefore I must necessarily ask, how much of it do you seriously intend within the next fifty years to be coal-pit, brickfield or quarry? ... from shore to shore the whole of the island is to be set as thick with chimneys as the masts stand in the docks of Liverpool: and there shall be no meadows in it; no trees; no gardens; ... that the smoke having rendered the light of the sun unserviceable, you work always by the light of your own gas: that no acre of English ground shall be without its shaft and its engine; and therefore no spot of English ground left, on which shall be possible to stand without a definite and calculable chance of being blown off it, at any moment, into small pieces.[3]

The words are self-revelatory. The industry has blighted the beautiful green; the smoke and the resultant darkness which rendered the sunlight unserviceable speak not just of the hellish possession of 'the sweet especial scene' by the satanic mills, they also embody man's spiritual wreck. Hopkins's poem "The Sea and the Skylark" aptly conveys the condition of the 'sordid turbid time':

> How these two shame this shallow and frail town!
> How ring right out our sordid turbid time,
> Being pure! We, life's pride and cared-for crown,
> Have lost that cheer and charm of earth's past prime:
> Our make and making break, are breaking, down
> To man's last dust, drain fast towards man's first slime.
>
> (ll, 9–14)

The lull and roar of the sea waves and the skylark's music put to shame man's 'shallow and frail town.' The town here refers to the resort town Rhyl in North Wales. The two sounds shut out from the speaker's experience of the 'sordid turbid time.' They give us a vision of happiness to be found in the world of nature but men immersed in materialism cannot respond to this. The natural world represented by the sea and the skylark in the octave are contrasted with the world of men in the sestet. The sestet is a condemnation of the human intrusion which has made the present world lose 'the cheer and charm of earth's past prime.' Words like 'shallow,' 'frail,' 'sordid,' 'turbid' speak of the ugliness of industrial towns and also convey a sense of decay and pollution. The last line refers to man's spiritual degeneration and hints at a reversal of Darwin's theory of evolution as it illustrates the complete regression of mankind back to inanimate matter or 'slime.'

Before I conclude, I would like to draw attention to another of Hopkins's lyrics 'Inversnaid.' In 1881, Hopkins became an assistant pastor at a Jesuit church in Glasgow, Scotland. While he was there he visited Inversnaid, a village located on the east bank of Loch Lomond in Scotland. The focus of the poem is on a mountain stream rushing down the hillside and emptying itself into the lake. The first three stanzas pictorially describe the different parts of the highland stream and its untameable force. The final stanza is an environmental plea to save the natural wildness and even today nothing can be more appealing than the lines which plead for conservation of backwoods:

> What would the world be, once bereft
> Of wet and wildness? Let them be left,
> O let them be left, wildness and wet;
> Long live the weeds and the wilderness yet.
>
> ("Inversnaid" ll, 13–16)

The poems of Hopkins definitely confirm, as observed by John Parham, that he achieved 'a fully recognisable Victorian ecology.' Many of his poems focus on the interactions between the physical, biological, and chemical environment. Such poems speak of the need to build a sustainable existence on this planet by dwelling on the interaction between humans, animals, and the environment with specific reference to the effects of modern technological advances. I think it would not be an exaggeration to say that reading Hopkins's poems and his journal entries can help us gain an understanding of the role ecology plays in conservation, wetland, and natural resource management, and forestry. As I conclude, I would like to refer to the words of Brian J. Day:

> His framing of ecological concerns in densely theological, moral, and spiritual terms marks him out as eminently Victorian. But at the heart of this spiritual ecology is the belief that all undefiled nature, rightly perceived, is sacred – or, as we might say in more secular terms, possesses intrinsic value – and that what we do to it we do to ourselves.
>
> (190)

Notes

1 Eric Thomson, who writes, 'The miscreant Fellows of Balliol are philosopher Thomas Hill Green (1836–1882) and ancient historian William Lambert Newman (1834–1923).'
2 Ibid.
3 See https://en.wikisource.org/wiki/The_Two_Paths/Lecture_III

Works Cited

Bate, Jonathan Ed. *Romantic Ecology: Wordsworth and the Environmental Tradition*, 1991.

Bump, Jerome "Hopkins, the Humanities, and the Environment" *Georgia Review.* 28.2 (Summer, 1974): 227–224.

Day, Brian J. "Hopkins' Spiritual Ecology in 'Binsey Poplars.'" *Victorian Poetry*. 42.2 (Summer, 2004): West Virginia University Press, pp. 181–194.

Hopkins, Gerard Manley. *The Correspondence of Gerard Manley Hopkins and Richard Watson Dixon*, Ed. Claude Colleer Abbott, 2nd ed. 1955; rpt. London: Oxford University Press, 1970.

———. *Further Letters of Gerard Manley Hopkins: Including his Correspondence with Coventry Patmore*. 2nd ed. London: Oxford University Press, 1956.

——— *The Journals and Papers of Gerard Manley Hopkins*, Ed. Humphry House. Compiled by Graham Storey, 1959; rpt. London: Oxford University Press, 1966.

——— *The Letters of Gerard Manley Hopkins to Robert Bridges*. Ed. Claude Colleer Abbott, 2nd ed., 1955: rpt. London: Oxford University Press, 1970.

——— *Sermons and Devotional Writings of Gerard Manley Hopkins*. Ed. Christopher Devlin. London: Oxford University Press, 1959.

Leavis, F.R. *New Bearings in English Poetry: A Study of the Contemporary Situation*. London: Penguin Books in association with Chatto & Windus, 1932.

Nixon, Jude V. "The Kindly Light: A Reappraisal of the Influence of Newman on Hopkins." *Texas Studies in Literature and Language*. 31.1 (1989): University of Texas Press, pp 105–142.

Parham, John Ed. *The Environmental Tradition in English Literature*. Burlington, Vermont: Ashgate, 2002.

Parham, John. *Green Man Hopkins: Poetry and the Victorian Ecological Imagination*. New York: Rodopi, 2010.

9 Statelessness, Environmental Refugee, and '*The Law of Humanity*'

Reading Hannah Arendt, Lawrence Buell, and Amitav Ghosh Together

Sajalkumar Bhattacharya

I

As I set myself to the task of writing this chapter, the Rohingya crisis, now internationally infamous, continues to worsen with more and more Rohingyas attempting to flee Rakhine State of Myanmar, and India and Bangladesh stubbornly continuing to harden their stands on this issue. Denied citizenship by Myanmar under the 1982 Citizenship Law, the Rohingyas constitute one of the largest stateless populations in the world today. Yet the Rohingyas are not the only examples of stateless people on earth. The number of stateless refugees is on the rise daily and continues to haunt the global academia, the human rights activists, the statesmen, and the politicians, each group in a different way. And it is precisely in this context that Hannah Arendt and her *The Origins of Totalitarianism* (1958) become more and more relevant. Hannah's succinct and incisive analysis of the concept of statelessness in this book has, of late, forced human rights activists to reconsider their issues at hand considerably, for Hannah's thoughts have problematized the discourse of human rights in such complex ways, as the very idea of human rights today is under the threat of being reduced to a fantastic, impractical one in various cases. Though Hannah's observations are based on the chain of events that followed World War I and a few decades thereafter, they find relevance globally every day now, for the problem of statelessness and the global nation-states' matching indifference to it that stunned Hannah way back in 1958 have only assumed gigantic proportions in the recent years. One of the major international problems that all developed nations (and even the developing ones) handle these days with acute discomfort is that of an unmanageably huge number of illegal migrants. Be it the Bengal borderlands or the American shores, the picture of a group of hapless young and old men and women, having no other identity to declare at the check posts of civilized nations than that they are living creatures, is as common today as the twigs of a tree. Political consciousness, administrative compulsions, and practicalities of national sovereignty have taught the civilized world to be utterly indifferent to these people, because they belong, literally, to nowhere. They have no legitimate claim, whatsoever, to any

DOI: 10.4324/9781003383970-12

identity that is the exclusive right of one who is within (and not outside) the geopolitical borders. Yet their problem, as Hannah succinctly points out in chapter 9 of *The Origins of Totalitarianism* (1958), is not their loss of homeland but their inability to find a new one on this earth. Consequently, 'once they had left their homeland, they remained homeless, once they had left their state, they became stateless, once they had been deprived of their human rights, they were rightless, *the scum of earth*' (Arendt, 267, emphasis added).

It would not perhaps be out of context to delve a little at this point into the history of this inhuman condition that Hannah details in the same chapter, for that will help us to gauge the depth of today's problem. Referring to the Peace Treaties and declarations by nations keen on enlisting their names in the League of Nations after the Paris Peace Conference in 1919, Hannah draws our attention to the shrewd fallacy underlying the sugar-coated promises therein. The treaties readily offered a host of basic rights to all legal inhabitants of the nation, irrespective of language, race, and religion. The rights of the minorities also stood protected. But what remained mutely but firmly implied in the day-to-day practical working system of the nation-states was that only the *nationals, and none but the nationals* would come under the protection of these rights. This excluded not only the non-nationals but also all those who had been divorced from the homeland and yet not been accepted in the host land. The legacy continues to date and the number of such people as these who are standing on the fringes today is astronomical: UNHCR, the UN Refugee Agency, puts it to more than 10 million of which one-third are children.[1] This number is on the rise continually, making statelessness the 'newest mass phenomenon in contemporary history, and the existence of an ever-growing new people comprised of stateless persons, the most symptomatic group in contemporary politics' (277).

The Rights of Man in 1789 promised to consider a human being important irrespective of the geopolitical borders that contain him, delivering him from the throes of the nation-state. But eventually, over the years, it turned out to be too idealistic to be considered for practical implementation. The nation-state, for all practical purposes, has emerged as a source of absolute power, and all ideas that it conceives and endeavours to put to practice are some way or other, steps towards consolidation of power and absolute control over its acknowledged citizens. So, in reality, the rights have never existed independent of governments. Consequently, the moment a human being is disowned by a government and rendered stateless, he finds himself abysmally alone, with no authority to protect his interest. At the end of the day, the much-hyped human right is reduced to a set of ludicrous, ineffective ideals, repeated reference to which makes a political speech grand, but which is never meant to be taken seriously. Rightless people do not have any guarantee of physical safety, because no government is bound to provide them; neither do they enjoy any freedom of movement as they do not have any address to move to or from (which even a jailed criminal has), nor any freedom of opinion, because no state seems to be bothered about what they say. Eventually, with all their physical, 'human' existents, the stateless people are non-existents on this earth:

> The calamity of the rightless is not that they are deprived of life, liberty, and the pursuit of happiness, or of equality before the law and freedom of opinion … – *but that they no longer belong to any community whatsoever*. Their plight is not that they are not equal before the law, but that no law exists for them; not that they are oppressed but that nobody wants even to oppress them.
>
> (295–296, emphasis added)

Yet, the whole problem is laden with an irony that makes it all the more despicable – 'The trouble is that,' as Hannah Arendt in The Origins of Totalitarianism points out,

> their calamity arose not from any lack of civilization, backwardness, or mere tyranny, but, on the contrary, that it could not be repaired, because there was no longer any 'uncivilized' spot on earth, because whether we like it or not we have really started to live in One World.
>
> (Arendt, 1958, 297)

This, therefore, is the crux of the problem: we are no more barbarians, but 'civilized,' and hence have no 'untidy' corner left within our society where to find some space for persons who have no geopolitical tags. And even within this neatly structured 'civilized' space (society), the only bond that connects one human being with another is that of power. It is this aggressive power in human beings that have engaged him in a conflict with not only his fellow human beings but also with all other living objects in nature. In the process, human beings have, in the first place, alienated themselves from Nature, and second, gravitated severally to different power groups rather than exist, as Humanity. These different power groups have taken shape as different nations, among many other factions, and have resisted any other sphere called Humanity to rise above it. Failure of Humanity, as a sphere, to exist above the nations has been, therefore, the main stumbling block to human rights. Ultimately human rights can only be safeguarded effectively by humanity and not by any nation-state. So, being a human and only a human does not, in this civilized world, guarantee human rights – one has to be a citizen in pen and paper. This perhaps is the deadliest evil that civilization has ever produced.

II

After more than 60 years of publication, the relevance of *The Origins of Totalitarianism* has received yet another dimension today because the concept of statelessness is no more restricted to the subject of geopolitical borders, but points, at present, to an immensely larger crisis that spans the entire planet. In this section, I would like to briefly assess this new dimension in the discourse of statelessness and how it is at crossroads with Lawrence Buell (talking of ecoterrorism) and John Barry (talking of resistance citizenship). Ecoterrorism is perhaps the most recent threat to the living world. Lawrence Buell, in his article "What Is Ecoterrorism" (2009), points to the politicized reading of the epithet 'ecoterrorism' by the Leftists and the Rightists, and how the term has been used by each of the two groups, in its way, to attack either of the stakeholders the epithet entails. While one group uses the term to level charges against the human intruders into the non-human world (e.g., the poachers) and talk at length about the threat they pose to the ecological structure, the other targets the eco-activists themselves as eco-terrorists, for their over-enthusiasm in eco-activism is often underlined by brutal indifference to human rights. The fact remains that in whatever way it is defined, ecoterrorism today is producing environmental refugees globally and rendering them stateless. The crux of the problem lies in its enormity, in its all-engulfing nature, because when we talk about environmental refugees, we talk not only about the human refugees, but the non-human living refugees – the flora and the fauna – as well. And importantly, as stateless entities, all these refugees are utterly deprived of human rights or animal rights, as the case may be. The irony Hannah points out in her essay, that (civilized) human beings themselves stand as resistance to human rights, is applicable in this broader canvas as well: while as Anthropocene, the human world is accelerating the process of ecological derangement with unprecedented madness every day, it cannot escape the unthinkable results of this atrocity. The human world thus is both the cause and the victim of this devastating terrorism. It is a known fact now that insatiable human greed and the resulting thoughtless exploitation of natural resources are at the root of an alarming ecological imbalance all over the world. But greed has insulated the human faculty of reasoning so incorrigibly that it has remained surprisingly oblivious to all warnings of Nature. The consequence is an unprecedented rise in the number of climate-related disasters and a consequent rise in the number of climate refugees – both human and non-human. Of all displacements worldwide today, the maximum number of cases is caused by climate disasters. Since 2008, an average of 24 million people[2] (data were taken from Internal Displacement Monitoring Centre, IDMC) has been displaced by catastrophic weather disasters annually. As climate change invites more storms, droughts, and floods in near future, climate scientists and migration experts expect the number to rise alarmingly. In Bangladesh, thousands of people are routinely uprooted by coastal flooding and rendered climate refugees. While many of them end in filthy slums with sub-human living conditions in and around Dhaka, the capital, the young and ambitious engage in a tortuous struggle to make it to India, the Arab countries, or the West. More often than not, these journeys end up, not in satisfactory resettlement, but perpetual deterritorialization. Bangladesh is largely clueless as to how to tackle the situation. India too has remained largely indifferent to date to the threats. While on the one hand, Kerala was badly affected by unusual monsoon rains in 2018, claiming at least 500 lives; Odisha reeled under the disastrous fangs of super cyclone Fani in May 2019; Chennai, on the other hand, faced the worst water crisis so far in July 2019. Only a few months back, West Bengal has been horribly affected by another super-cyclone Amphan, and it will take months to recover from the shock and the calamitous consequences. The report from NITI Aayog – a government think tank – has warned that at least 21 Indian cities will run out of groundwater by 2020.[3] Many islands of the Sundarbans have also started to disappear under the rising sea level. Again, the Amazon rain forest burnt for almost a fortnight recently. And ultimately, all these are causing large-scale human and non-human displacements, directly or indirectly. Available evidence indicates that climate change and environmental degradation have crucially affected mass human (and non-human) mobility internationally. There are, right now, almost more than 20 million climate refugees globally,[4] and no government across the globe still have any idea regarding what to do with them. Available statistics of non-human climate refugees would be equally alarming. For example, the number of Southern Cassowary, a large, flightless bird species, third in size after the emu and the ostrich, and an important seed disperser of the rainforest trees, has been severely affected by the Cyclones Larry and Yasi in 2006 and 2011. Many older cassowaries perished and the younger ones were forced to migrate to the adjacent suburban areas and tourist resorts. There they died in large numbers due to starvation, traffic accidents, and attacks by dogs.[5] As WWF reports, Pacific walruses are under serious threat from loss of habitat caused by climate change. With the steady decline in the sea ice throughout the Arctic, more and more walruses are forced to seek refuge ashore and travel long distances up to 250 miles round trip to reach their food supply. This phenomenon, known as a 'haulout,' is making them more and more endangered as a species, forcing them to try to adapt forcefully with unconducive living conditions.[6]

What is the solution then? Would one sit an idle, passive witness to the impending disaster? Or should one think along some positive line and engage in an incessant active endeavour to rehabilitate the stateless human and non-human world to keep this planet of ours thriving as long as one can? Arendt has unambiguously expressed her distrust in state laws and would rather invest her faith in 'a new law on earth, whose validity this time must comprehend the whole law of humanity' (Arendt, Preface to the First Edition, IX). John Barry, in his article "Resistance Is Fertile: From Environmental to Sustainability Citizenship" (2005), proposes something more concrete and aggressive through his term 'Resistance Citizenship.' Talking about the need for 'critical citizens,' and not merely passive law-abiding citizens, to strive for the achievement of a sustainable society, Barry has reiterated in this article, as also in his other works, the need to engage oneself in 'corrective or oppositional work in the form of resistance' (32) to ensure sustainable development in both the human and the non-human worlds:

> If we accept the claim that the demands of justice include action and often *oppositional action to rectify injustice*, then a strong connection can be made between justifying resistance citizenship in the cause of sustainable development. This is the case if we view sustainable development as denoting justice between generations, between fellow citizens, and between different parts of the world – intergenerational, social, and global justice, respectively. In this case, expressing the demands of sustainable development in the language of justice strengthens the case for resistance forms of citizen action to create a less unjust and unsustainable society both locally and globally.
>
> (44, emphasis added)

Interestingly, and more relevantly for us in this essay, Barry offers us a suggestion (though not without a note of personal doubt) that this may apply to the case of the non-human world as well:

> Although more contentious, there is also the argument that we can include relations between humans and animals within the remit of justice, and this is certainly the prevailing rationale among sections of the animal rights/welfare movement.
>
> (44)

On the whole, therefore, both Arendt and Barry express considerable doubt over the efficacy of state laws towards the achievement of holistic welfare of the living world and would rather encourage to think beyond the institutional borders.

III

As members of a Department of Literature, one crucially relevant point for us at this juncture might be to examine the role of Literature *per se* in contemporary eco-activism to save our planet: is it merely to register passively the devastation wrecked by a cyclone or a flood as backdrops of tales of eternal human emotions, conquest, or suffering? Or does Literature have a more active responsibility to discharge in this regard? For unless we justify our position convincingly in this project, and by 'our,' I mean all of us who are associated with literary productions as authors or readers, we shall continue to be slighted as being useless intellectual amateurs who draw sustenance from these debates, but are removed light-years from the ground reality of eco-activism. I find Amitav Ghosh particularly relevant here, for he seems to have already bothered himself with this question on our behalf and has confirmed our boarding passes to the ship of eco-activists of today.

Ghosh's recent works (both fictional and non-fictional) not only encapsulate in their way the points Arendt and Barry have raised and popularized, but also seem to complete the argument by picking up relevant micro-narratives from human history, either tracking down factual details or presenting them under the garb of fiction. And, in the process, he makes the participation of a littérateur like him not merely relevant, but almost unavoidable in this project. I particularly have in mind his seminal essay "Wild Fictions" (2008) and his recent non-fictional work *The Great Derangement* (2016) in this context.

In "Wild Fictions," an essay which has surprisingly not received its due critical attention so far, Ghosh makes some observations that neatly build up an argument advocating the importance of relevant narratives, real and fictional, to address the ecological problems of today's world. To begin with, Ghosh unambiguously warns that to take an exclusivist approach to save the ecological balance in today's world would be futile, if not suicidal: if the government policies, in their new-found over-enthusiasm in saving the non-human world, segregate the human component in it, and end up in producing large groups of uprooted refugees, it would be disastrous, for this approach has always 'been harmful not just for the ecosystem people' but also for the 'very environment it sought to protect' (WF, 16). Dreaming of saving the ecosystem, thus, would be a 'wild fiction,' in the truest sense of the term. Taking a cue from Ramachandra Guha, Ghosh, in this essay, points to the most basic issue of ecoterrorism – if one has to think of the environment, then one has to put both the human and the non-human components together. Emphasis on one over the other disturbs the balance disastrously.

The crux of the problem lies in the general, traditional attempt at trying to understand ecology by assuming that its two components – human and non-human – are locked in a binary relation. That has prodded us for a long time either to support this component or the other and to rack our brains how one group has victimized the other. I still recollect one *adda* in my undergraduate hostel life that proved to be a lesson of a lifetime for me. One afternoon, having nothing better to engage ourselves in, some of us raised a storm in the tea-cup over a piece of news published that day in the newspaper about the killing of a tiger by a group of villagers in the Sundarbans. We, 'urban educated elites' of Kolkata, were furious that the insensitive villagers had killed an invaluable Royal Bengal tiger, after trapping it in somebody's kitchen. We were stunned at the irresponsibility of the 'insensitive' villagers, clear from their failure to realize the importance of Tiger Conservation Projects in the Sundarbans and groped for words that would amply give vent to our anger. Eventually, there was one boy among us from the remote Gosaba area of the Sundarbans, an area which was frequented by tigers. He was listening to this heated discussion all the while but choosing to keep silent. Suddenly he could restrain himself no more and started shouting at us. He found us to be fools and selfish and invited us, with a note of bitter sarcasm which he hardly tried to conceal, to spend a few nights with him in his village and see whether we still retain the same concern for our 'favourite' Royal Bengals. After cooling down a bit later on, he tried to impress on as how it is one thing to write 'learned' articles on conservation of tigers, and a completely different experience to lie on a sleepless bed in a tiger-infested village without basic amenities of life scarcely available; how a chill suddenly runs down one's spine when one suddenly feels the urge to relieve oneself in the dead of the night, because there is an eerie feeling, a panic, that tells one almost with a note of fateful certainty that the moment one opens the door, one may face the fearful Royal Bengal or a king cobra, lurking in the dark for hours for a human body. Our friend grew passionate while he tried his best to make us understand how different it feels to share the same geopolitical space with a tiger. The point he was underscoring throughout was that the government, and even more unfortunately the urban elite society, is all sympathy for the tiger and is least interested in protecting the human habitation or empathizing with the utter helplessness of the human world at Sundarbans. He tried to strip us of our amateurish attitude to the whole affair by pointing out the chasm between reality and our academic exclusivity. A note of sheer anguish was clear in his voice all the while.

Against this, I would like to juxtapose a micro-narrative that Ghosh retrieves from a footnote of an essay documenting a real event of July 1852 on the banks of river Hooghly and narrativizes in "Wild Fictions": on a hot July day in Kolkata, the river Hooghly flowed over its embankment and after the water receded, a school of gigantic creatures was spotted on the wetland. The news spread and an Englishman Edward Blyth by name came running. He was then the curator of Natural History at the Asiatic Society. He spotted some 20 huge whales which were in great distress. Blyth was surprised that a large crowd had gathered but not a single whale was killed for meat, as he had expected. Instead, this crowd had been trying to rescue the animals and some had already been saved. Blyth selected some and secured them with ropes to dissect them for research purposes and made arrangements so that they were not let loose by the 'foolish' natives before he returned the next day. But to his utter surprise, when he came back the next day, he found these creatures had been rescued, and the villagers would go even to the length of incurring the wrath of a sahib to free them. One pragmatic reason was of course that it was a local fishing ground and the fishes had to be saved from the whales. 'Yet,' in 'Wild Fictions', Ghosh says,

> compelling as these pragmatic reasons might be, I find it hard to believe that they were not allied also to a certain sense of awe, wonder, and even compassion at the sight of the distress of these majestic creatures.

> Is it possible that there was no talk among the villagers of divine visitations, no stories told of signs from the heavens?
>
> (Ghosh, 1979, 6)

Ghosh continues,

> such emotions might appear to have little in common with an ecological awareness, but if indeed there is, in cultures at large, as well as works of literature, then surely it would consist in an overlapping of the pragmatic and the poetic, a broad acknowledgment of mutual dependence, in which rights, mutual obligations and a sense of wonder are seamlessly merged.
>
> (Ghosh, 1979, 7)

These two incidents, considered together, would point to the error that has undermined even the most sincere human efforts at eco-conservation: the human and the non-human components have been considered as mutually exclusive components in the problem of conservation, engendering the threat of eco-terrorism. For almost two centuries, efforts have been directed either at making the planet more habitable for humans necessitating ruthless destruction of the non-human wild world or criminalizing the human beings for their 'intrusion' in the wilds. In the nineteenth century, Ghosh points out, the generally accepted view among ('academically trained') foresters was that 'the presence of people was always a threat and never an asset to forests', and hence if forests have to be saved, human encroachers have to be chased out. In the hay-days of the British Raj in India, the government enticed all concerned to believe that the 'backward' local population was the most dangerous threat to animals. Therefore, the forests needed to be 'disinfected' from the human element. And shockingly enough, the same colonial perception to date continues to inform the forest departments in India at large, resulting in the alarming increase in the number of environmental refugees in every reserve forest. Citing two glaring examples from Rajasthan and Madhya Pradesh in this context, Ghosh laments that, 'in effect, over many decades, there has been a kind of "ethnic cleansing" of India's forests: indigenous groups have been evicted or marginalized and hotel chains and urban tourists have moved in' (np), so that 'the *administrative apparatus of eviction*, restriction and so on ... make *these wildernesses conform to the tourist's notion of the "pristine"*' (np, emphasis added). Consequences have been equally disastrous for both the human and the non-human world.[7]

Maybe things are changing very slowly of late, but overall, be it on the level of policy formation or legislation, or even in the human psyche, the concept of sustainable conservation keeping all the components of the environment in a robust connection with each other as opposed to single-species conservation is yet to gain ground satisfactorily. More awareness (in the sense that our attitude to Nature as a whole undergoes thorough change) is necessary, and that too, immediately. And it is precisely here that Ghosh reiterates the role of stories. According to Ghosh, Fiction has an extraordinary power to shape our ideas of nature, because 'if there is anything distinctive about human beings, as a species, it consists, ... in our ability to experience the world through stories' (np).

As Ghosh shows us in this essay, there has been a global evolution down the ages in the conceptualization of Nature in fictional/philosophical works. Saint-Pierre in the late eighteenth century, along with his mentor Rousseau and of course the Romantics of the nineteenth century, must be given their due honour in the history of environmental activism. Saint-Pierre in particular can be held as one of the initiators of this activism. But in their over-enthusiasm in establishing the sacredness of Nature, they have taken such an exclusivist approach that the human element begins to be considered a threat in this project and hence needs to be resisted actively or even aggressively. So, Pierre's narrative "The Indian Hut" (1791) or his novel *Paul et Virginie* (1788) now stands dated. As Ghosh sarcastically puts it, these great worshippers of Nature did not have to earn their living from the woods, after all!

Even the Bonbibi myth has its limitations, Ghosh points out. It has of course done its bit by drawing attention to human greed and its disastrous effects and by reiterating the need to maintain an equitable dispensation between the wilderness ruled by the tiger and the human settlement. While the importance of the myth can hardly be overestimated, the fact remains that its limitations are exposed in today's context, mainly because while the legend works out in a local context, it could not foresee the disruptions the contemporary world faces today. For example, with the mushroom growth of commercial fisheries in this area, fishermen are using nets of very fine mesh to catch the tiny spawn of prawns. This has been a recent threat to aquatic life and a devastating one, but the Bonbibi myth has no answer for it.

So the point is, new stories should be written keeping in mind the new, emerging threats. Why stories? How does it score a point over sciences that have remained engaged in conserving ecology in their way? Ghosh opines that natural sciences like Zoology, Botany, and Geology are equally biased for these sciences 'direct a gaze of concentrated interpretive scrutiny towards the curtain of signs that is called "data"'. In the process, as Ghosh rightly observes, 'Science cannot be the final arbiter in the matter of our relationship with Nature' (np), for its methods cannot 'address questions of meaning, intention and lived history' (np), because 'the conditions of scientific inquiry are such as to require a radical separation between the inquirer and the field of study'. To be objective, it destroys the subjective human element. Scientists try to create the condition of a laboratory within inhabited landscapes. Then Ghosh looks at the domain of politics because human action and subjectivity come under its purview. But alas, 'the limitation of political action ... is that it cannot generate the imaginative resources necessary to rethink the human relationship with nature' (np).

Ultimately, therefore, Ghosh invests his trust in fictional imaginings which alone can grasp the vast spectrum of experience as the human mind is capable of conceiving. A poet is more equipped than a historian only because he can transcend the borders of recorded history and can bring out the micro-stories that grand-narratives tend to iron out. Consequently, he is the only person to record the 'environmental uncanny' (TGD, 42)[8] in literary fiction. Such stories alone could have contributed meaningfully to eco-activism by grooming a unique kind of 'environmental unconscious,' to borrow a term from Larry Buell.[9] Therefore, Ghosh has lamented again and again in *The Great Derangement* the lack of such stories in today's academia:

> When the subject of climate change appears in ... publications, it is almost always in relation to non-fiction; *novels and short stories are very rarely to be glimpsed within this horizon.* Indeed, it could even be said that fiction that deals with climate change is almost by definition not of the kind that is taken seriously by serious literary journals: ... It is as though in the literary imagination climate change were somehow akin to extraterrestrials or interplanetary travel.
>
> (9–10, emphasis added)

This, according to Ghosh, is a pity, because only the faculty of literary imagination can transcend the limitations of a journalistic report of what has *already happened* and can predict the unthinkable disaster that may befall the living world in near future. To 'think' of the 'unthinkable' beforehand is rare faculty of literary imagination, which also engenders in a poet his responsibility towards the universe – a responsibility which can hardly ever be shirked. This 'thinking,' as Ghosh likes to put it, does a two-fold job. While it goes a long way in nurturing the 'environmental unconscious' discussed earlier, warning the world of the impending disaster and even bracing it for fatal shocks from the environment, it also provides the only possible strategy to clear a path of survival amidst the debris of eco-disaster: to establish the human 'kinship with other beings.' *The Great Derangement*, like most other works by Ghosh, ends on a note of hope, provided Literature does its bit to cement this kinship:

> The struggle for action will no doubt be difficult and hard-fought, ... (b)ut I would like to believe *that out of this struggle will be born a generation that will be able to look upon the world with clearer eyes than those that preceded it*, that they will be able to transcend the isolation in which humanity was entrapped in the time of its derangement; *that they will rediscover their kinship with other beings, and that this vision*, at once new and ancient, *will find expression in a transformed and renewed art and literature.*
>
> (216–217, emphasis added)

IV

Reference to at least two fictional works by Ghosh, *The Hungry Tide* (2005) and *The Gun Island* (2019), would be pertinent here, for they amply illustrate the points he raises in his essays discussed in the earlier section. Having dwelt at length exclusively on the issue of geopolitical borders, marked and sanctified by the state to consolidate hegemonic power, Ghosh has, in recent times, engaged himself in the exploration of recent global environmental crisis and its effects on a different set of borders: that between the human and the non-human world. As an anthropologist, he has all along been touched by human displacement, and the recent alarming rise in the number of climate refugees globally along with the increasing rapture between the human and the non-human world has prodded him to explore the art of fiction to warn the world of the impending 'great derangement' on the one hand and forge the much-needed kinship – kinship between the human and non-human world, and among the groups and subgroups created within the human society through race, nationality, and power.

The Hungry Tide raises two crucially important issues. Playing out a group of urban characters in the locale of the remote Sundarbans, Ghosh at first explores the possibilities of effective communication between two antonymous human worlds. Charting the transformation of Nirmal and Nilima, Kanai, and Piya, Ghosh posits this possibility in civic facilities created effectively by 'cosmopolitan'[10] individuals. Close human relationships inspired by altruistic love alone can break the borders, Ghosh suggests. This alone can consolidate a kinship bond among various unequal standings. I will only pick up the cases of Nilima and Piya to illustrate my point.

When Nilima sets her feet on the sticky mud of the Sundarbans along with her husband, she was a complete stranger to this land of abject poverty and unspeakable misery, even when it was barely a hundred kilometres away from her affluent Kolkata home. But once she settles down there with firm determination, she engages herself in forging meaningful human bonds with the hapless fisherfolk there. This gradually evolves into the Badabon Trust, Nilima becoming over the years, the caring mother-figure of the entire community, which needed her to care for its well-being. Acting as a cohesive force, she creates an effective family around her, irrespective of all borderlines, threatening to segregate one from the other.

Piya is an even more classic example of the formation of this kinship bond. She comes to this place as a researcher, an ambassador from the Western academia as it were, and to whom the local culture, mode of living, and even the language is foreign. Still, with the passage of time, she overcomes all challenges to establish an effective, deep communication both with the mute world of the Orcaella and the marginalized, hapless human community of the Sundarbans. The mutual attraction and the mute communication that develops between Piya and Fokir in the course of the novel is an enactment of meaningful, deep communication between the privileged, urban elite and the non-privileged, rural, marginalized societies. They represent antonymous social, economic, and cultural positions and do not even follow each other's language, but that cannot stop them from forging a familial bond between them even in the moment of imminent death. This bond transcends death, because the project, in which Piya engages herself after the cyclone kills Fokir, is her spontaneous attempt to perpetuate his memory. Ghosh's attempt at sensitizing his readers to the deep communication that Fokir could develop with the Orcaella dolphins, the Royal Bengal tiger, and the natural elements in the novel, along with the empathetic deep communication he could develop with Piya, illustrates precisely what he (Ghosh) proposes as the function of Fiction in a world shaken by the disruptive forces of ecoterrorism. Seamlessly integrated to this is the incident of Morichjhapi in this novel, which itself is a struggle born out of the will of thousands of uprooted refugees to survive against all odds. Their aim was 'quite straight forward. They just wanted a little land to settle on. But for that they were willing to pit themselves against the government. They were prepared to resist until the end' (HT119). This indomitable common will to survive against all odds strengthens the kinship bond within them and make them all members of one *bastuhara* (homeless) family. It is as if a family that they declare their common identity – '*amrakara*? – *bastuhara*' (254) (we are the homeless – that is our identity), and it is as a family that they fight till the end. By bringing these two stories together, Ghosh creates in *The Hungry Tide*, an empathetic awareness of one and all regarding the pressing demand of deep communication among all segregated entities – human and non-human, all stateless refugees – political or climate. Standing on the verge of disaster, Ghosh earnestly feels the urgency of this awareness, if one is to envisage a solution to the problem of statelessness. That is why in "Wild Fictions," he promotes the fictional works of novelists like Mahasweta Devi and Gopinath Mohanty.

Finally, in *The Gun Island* (2019), Ghosh examines the problem on a larger, global canvas. Narrated by a Brooklyn-based NRI, Dinu, the novel is a sequel to *The Hungry Tide*, bringing together the Sundarbans, Los Angeles, and Venice and indicating how the threat of statelessness, political as well as ecological, links all these apparently distant geopolitical locations. *The Gun Island*, therefore, is not a once-upon-a-time story, but very contemporary, underlining for us the glaring reality of today.

A wonderful storyteller, Ghosh has woven into the narrative quite a few stories together, enmeshing them inextricably to show the unavoidable interconnectedness of all incidents of human and non-human displacements – temporally and spatially. Tipu and Rafi represent the underprivileged third-world young generation, keen on freeing themselves from the misery of their nation and landing up in the dream world of the West. Desperate on snatching their share of the privileges denied to them for decades, they now bother no more for passports or visas to cross borders frequently, but illegally. Tipu becomes a middleman in human trafficking, convinced that by helping people move out of the Sundarbans, he is offering them an 'essential service':

> What I'm doing is I'm offering an essential service. In these parts, there's a whole bunch of dirt-poor, illiterate people scratching out a living by fishing or farming or going into the jungle to collect bamboo and honey. Or at least that's what they used to do. But now the fish catch is down, the land's turning is salty, and you can't go into the jungle without bribing the forest guards. On top of that every other year you get hit by a storm that blows everything to pieces. So what are people supposed to do? ... If you've got any sense you'ld move and to do that you need someone who can help you find a way out.
>
> (60–61)

And out of this basic, instinctive need to live, they take the help of the modern internet. 'The internet is the migrants' magic carpet' (61), Tipu says, making the question of legality utterly irrelevant. In a sequence of bizarre events, Tipu, Rafi, and a host of other 'illegal' migrants travel from Bangladesh, through India, Pakistan, and Iran, to Turkish borders. Rafi can ultimately make it to Venice but Tipu gets stranded. And this human story of displacement runs parallel with an equally hazardous tale relating to the non-human world – how Piya's Orcaella dolphins have to move out of their natural habitat, the Sundarbans, and die in large numbers because of the polluting waste from refineries; how the yellow-bellied sea snakes are displaced from warmer waters, posing a threat to the human world in Venice beach as well as in Ventura Beach, California; and how the massive wildfires raging about Los Angeles for several days render a huge chunk of the human and non-human population, as environmental refugees.

As Ghosh, in the course of the novel, enquires into the phenomenon of large-scale production of refugees today globally, he comes up with some such observations that prove crucial in underlining the main theme of the present chapter. Remarkably, Ghosh locates the cause of this large-scale displacement – human and non-human – and the consequent forced migration, in a ruthless power structure. In the human world, the colonizers initiated this, centuries back by forcing indentured labour of the colonies to move to the colonized world for the benefits of the 'powerful' colonizers. Almost parallelly, the human world has colonized the non-human world by pushing all other less powerful creatures out of their natural habitat in the name of development, for centuries. Time has now come to settle the accounts and the tables are being turned. Spaces that have meticulously been sanitized to keep off the 'other' are now being trespassed, with the 'empire' invading back this space as their own. Perhaps this is how History and Nature restore balance. The conclusion of the novel is very significant in this context. In a masterstroke, Ghosh brings representatives of all displaced and disempowered species together – human beings, dolphins, and the migratory birds ('a series of migratory patterns intersecting in an unusual way,' as Ghosh puts it). Against this backdrop, we find a small, blue fishing boat, carrying some illegal migrants from different Otherized geopolitical spaces of the world, trying to reach the shores of Italy. The boat faces strong resistance from the right-wingers of Europe, who were creating tremendous pressure on the government to stop these intruders. This is basically because the Europeans have now realized with a sense of horror that their age-old 'civilising mission' and 'colonial project' have now backfired:

> The entire project had now been upended. The systems and technologies that had made those massive demographic interventions possible – ranging from armaments to the control of information – had now achieved escape velocity: they were no longer under anyone's control. This was why those angry young men were so afraid of that little blue fishing boat: through the prism of this vessel, they could glimpse the unravelling of a centuries-old project that had conferred vast privilege on them in relation to the rest of the world. In their hearts, they knew that their privileges could no longer be assured by the people and institutions they had once trusted to provide for them. The world had changed too much, too fast the systems that were in control now did not obey any human master; they followed their own imperatives, inscrutable as demons.
>
> (GI, 279–280)

Eventually, the boat becomes 'a symbol of everything that's going wrong with the world – inequality, climate change, capitalism, corruption, the arms trade, the oil industry' (199). The state (represented by the minister in the novel) bluntly refuses them entry. But the immigrants receive huge support from leftist groups and NGOs from within the country as also from countries like Germany, Singapore, Hungary, Russia, and Australia. Finally, the Admiral of the Navy, being led presumably by humanitarian obligations, violates the directive of his state and allows these migrants in. In the novel, he risks his high position and legal threats but declares that he is bound by a set of laws that transcends the law of the land:

> I have nothing to fear from the law ... *I have acted in accordance with the law of the sea, the law of humanity, and the law of God.* If I am tried, those are the laws that I will answer to.
>
> (Ghosh, 284–285, emphasis added)

This is where Ghosh, Arendt, and John Barry come together. And it is these kinds of stories that would prioritize the 'law of humanity' over any other law that Ghosh has been stressing the need of again and again. And only a poet can be entrusted with such a task, for only a poet, among all human beings, is richly endowed with an unshakable trust in the living world and Life. Therefore, he alone can dream of and can help all others dream of a planet where each living species will find a habitable space as its own. Utopian? – Yes, admittedly so. But when the planet is becoming more and more dystopic every day with the level of carbon monoxide rising to critical levels and the borders of the nation-state becoming tighter than a noose, who but a storyteller, a poet, can constantly keep the hope to stay alive with the Utopian mantra and make us strive to attain that state? That is why Ghosh laments the dearth of writers in contemporary times who can engage actively – imaginatively with the crisis of displacement. Hence is the need for as many storytellers as we need specialized eco-activists.

As I conclude this chapter, Covid-19 has invaded the world with unprecedented swiftness leaving it almost helpless. The number of infected persons throughout the globe is increasing astronomically. This leaves us with a sense of humility as living beings who have always considered them to be the most superior species. As we are slowly being prepared to assimilate the truth that we shall ultimately have to learn to live with the virus, we also realize how our anthropocentric activities have triggered a fatal retaliation from Nature at last. But even the Corona pandemic has a silver lining for all of us. This pandemic, as well as the massive deterritorialization of both humans and non-humans, is opening up new avenues for developing ecological consciousness. Ecocritics like Ursula Heise believe, even though deterritorialization uproots people from their place of birth and weakens the bonds of culture but this worldwide deterritorialization of both humans and non-humans imbibes in all living beings a sense of planet, rather than a sense of place. Therefore, Heise, in the conclusion of her book *Sense of Place and Sense of Planet: The Environmental Imagination of the Global* (2008), stresses the necessity of narratives that would 'strive to find effective aesthetic templates by means of which to convey … a dual vision of the Earth as a whole and of the different earths that are shaped by varying cultural contexts' (210). Such narratives, according to Heise, would participate in the search for 'images of a new kind of eco-cosmopolitan environmentalism that might be able effectively to engage with steadily increasing patterns of global connectivity, including those created by broadening risk scenarios' (210).

Interestingly, Heise's idea of developing a 'sense of planet' echoes Gayatri Chakravorty Spivak's concept of 'planetarity,' which she discusses in *Death of a Discipline* (2003). Spivak notes, 'If we imagine ourselves as planetary subjects rather than global agents, planetary creatures rather than global entities, alterity remains underived from us' (72). Thus, 'planetarity' is a state that demands a radical openness to pluralization and consideration of difference that exists beyond human ontologies. So, only an active engagement with the concept of 'planetarity' may imbibe in us the much-needed 'sense of planet' and lead us, consequently to put our faith in the 'law of humanity.' Unless we foster this feeling of togetherness in us, unless we realize the underlying interconnectedness among every living species across all sorts of borders, nothing can save us from extinction in near future. Amitav Ghosh has effectively engaged himself in this task of creating this awareness in his recent novels. In *The Gun Island*, he has repeatedly used the Sanskrit phrase *Sasagara Basumati* (The entire landmass and water bodies on Earth) to stress this idea of interconnectedness. We look forward to more littérateurs following suit globally.

Acknowledgement

I gratefully acknowledge my indebtedness to Ms Rumela Saha, integrated MPhil–PhD scholar working under my supervision, for her help in writing this chapter. She has helped me with several references and insightful suggestions.

Notes

1 UNHCR – United Nations High Commissioner of Refugees– was established in 1950, during the aftermath of World War II, to help millions of Europeans who had fled or lost their homes. Today, this is a global Refugee Agency, functioning for around 69 years, dedicated to saving lives, protecting rights, and building a better future for refugees, forcibly displaced communities, and stateless people, fleeing from violence, persecution, war, or disaster. https://www.unhcr.org

2 IDMC – Internal Displacement Monitoring Centre – is an independent body providing a source of data and analysis on internal displacement, based in Geneva. In 1998, IDMC was established as a part of the Norwegian Refugee Council (NRC). The purpose of IDMC is to provide high-quality data, analysis, and expertise on internal displacement to inform policy and operational decisions that can reduce the risk of future displacement and improve the lives of internally displaced people worldwide. https://www.internal-displacement.org/

3 NITI Aayog – The National Institution for Transforming India, also called NITI Aayog, is a policy think tank of the Government of India, formed via a resolution of the Union Cabinet on 1 January 2015, replacing the Planning Commission instituted in 1950. NITI Aayog aims to encourage the involvement and participation of globally reputed policymakers, experts, and administrators in India, to share their expertise and knowledge in policymaking and good governance with their Indian counterparts. NITI Aayog acts as a quintessential platform of the Government of India to bring States to act together in the national interest and thereby foster Cooperative Federalism. In a report titled *Composite Water Management Index: A Tool for Water Management* published on 14 June 2018, NITI Aayog claimed that 21 Indian cities would run out of their supply of groundwater by 2020.http://social.niti.gov.in/uploads/sample/water_index_report.pdf

4 The 2001 *World Disaster Report* of the Red Cross and Red Crescent Societies reported the estimate of 25 million current 'environmental refugees,' cited in International Organization for Migration Research Series, 2008, titled *Migration and Climate Change*. International Organization for Migration (IOM) is an intergovernmental organization that acts with its partners in the international community to assist in meeting the operational challenges of migration, advance understanding of migration issues, encourage social and economic development through migration, and uphold human dignity and well-being of migrants. https://www.iom.int/

5 Australian Government, Department of Environment and Heritage 2004, published a fact sheet on *Threatened Species and Communities: The Southern Cassowary*. This report mentions that 'the southern cassowary is one of Australia's most imposing birds — large, colourful, and flightless. It is found only in the dense tropical rainforests of north-east Queensland. Continuing clearing and fragmentation of rainforest, and increased mortality from cars and dogs have reduced cassowary numbers to perhaps as few as around 2000, threatening the species with extinction.' https://www.environment.gov.au/system/files/resources/a6d7361a-a140-4a94-9863-f52896e18774/files/cassowary.pdf

6 WWF – World Wildlife Fund, summer 2017 issue published the article "Walrus Habitation on the Edge," authored by Isabella Groc, drawing our attention towards the impact of climate change on Pacific Walrus. https://www.worldwildlife.org/magazine/issues/summer-2017/articles/walrus-habitat-on-the-edge

7 Ghosh has acknowledged his indebtedness to Ramachandra Guha, in particular, in this connection. So Ghosh's take on this issue should be read most profitably as an extension of Guha.

8 Ghosh in *The Great Derangement* argues that since fiction has all along remained engaged in capturing the uncanny, e.g., the ghost stories. It should now attempt at dealing with the 'environmental uncanny' – the uncanny events in today's universe triggered by climate change, in which both the human and the non-human world have contributed in large measures.

9 In his book *Writing for an Endangered World: Literature, Culture and Environment in the US and Beyond* (2001), Larry Buell defined this term as a 'residual capacity of individual humans, authors, texts, readers, communities to awake to a fuller apprehension of the physical environment and one's interdependence with it.' Ghosh has borrowed this term from Buell and used it in "Wild Fictions."

10 Amanda Anderson defines a cosmopolitan as a person operating at a 'reflective distance from [his] original and cultural affiliations, [and possessing] a broad understanding of other cultures and customs and a belief in universal humanity' (Anderson 63). Taking a cue from this, Terri Tomsky, in his essay "Amitav Ghosh's Anxious Witnessing and the Ethics of Action in *The Hungry Tide*," describes Nirmal, Nilima, Kanai, and Piya as cosmopolitans. Another interesting discussion of cosmopolitanism is there in Ania Spyra's 'Is Cosmopolitanism Not for Women? Migration in Qurratulain Hyder's *Sita Betrayed* and Amitav Ghosh's *The Shadow Lines*.' She says, 'Cosmopolitanism ... does not rest simply on global mobility, ... traveling between communities and nation-states, crossing boundaries in a relentless search for a space of belonging. *It is active belonging*, rather, *that I see as an important variable in my understanding of cosmopolitanism*' (np, emphasis added). It is this 'active belonging' that Ghosh illustrates through Nirmal, Nilima, and most importantly, Piya, that I want to highlight in my reading of the novel.

Works Cited

Anderson, Amanda. *The Powers of Distance: Cosmopolitanism and the Cultivation of Detachment*. Princeton University Press, 2001.

Arendt, Hannah. *The Origins of Totalitarianism*. The World Publishing Company, 1958.

Barry, John. *Resistance is Fertile: From Environmental to Sustainability Citizenship*. 2005. 153–166.

10 The Rhetoric of Space

Space and Human Behaviour in Nathaniel Hawthorne's "Young Goodman Brown" and *The Scarlet Letter*

Pradipta Sengupta

To add anything new to Hawthorne seems to be a daunting task, and to shed any new light on his masterpiece, *The Scarlet Letter*, seems to be an impossible one. Given the wide ken of research[1] being carried out in this masterpiece, it becomes really difficult to make an original contribution to research. Yet, my humble submission in this chapter is to explore some of the insights from space theory and try to explore Hawthorne's deep-rooted liaison with his environment, with specific references to "Young Goodman Brown" and *The Scarlet Letter*. With the cachet of his genius, Hawthorne makes ingenious use of 'space' in which culture, history, human psychology, and human behavioural patterns get wonderfully enmeshed. Exploring the varied ramifications of space, this chapter proposes to examine his use of space in his short story "Young Goodman Brown" and his masterpiece, *The Scarlet Letter*. My humble attempts in this chapter are to substantiate how space reflects human psychological traits, how certain spaces evoke certain behavioural patterns, and how the physical and the psychological are wonderfully enmeshed together in Hawthorne's unique treatment of space in these texts.

Regional specificity is abundantly found in American literature. Topography has drawn the attention of many American writers. As Robert Jackson has so pertinently argued in his wonderful research, *Seeking the Region in American Literature and Culture*:

> Concepts of the region are particularly important in the American context not only because of the region's role as a primary repository for the kind of highly complex culture produced by such a nation but also because the American region stands in distinct contrast to European notions of region and regional identity.
>
> (4)

Thus, space became one of the major concerns of the Europeans who had settled down in this new land and were ultimately transmuted into 'Americans.' The very fact that Americans are virtually displaced Europeans made them more amenable to regional specificity. For example,

DOI: 10.4324/9781003383970-13

Robert Frost deals with New England scenes, while William Faulkner is interested in the American South. While Hawthorne delineates the annals and stories, mostly veering around his birthplace Salem, the same holds of John Updike, who, in the early phase of his prolific literary career, offers vivid glimpses of American middle-class suburbia from Shillington, Pennsylvania. One may go on adducing more examples from American literature to substantiate the penchant of certain writers for certain regions and spaces: Washington Irving, Tom Wolfe, and E.L. Doctorow for New York; Cormac McCarthy for the United States Southwest; Stephen King for rural Maine; Reynolds Price and Jim McCorkle for North Carolina; Pat Conroy for South Carolina; Louise Erdrich for North Dakota; Stephen King for rural Maine; Bobbie Ann Mason for Kentucky; John Nichols for Mexico; Scott Turow for Chicago; Patricia Cornwell for Richmond; among others. Long before the advent of theory, space, place, and region had fascinated writers belonging to every age, and Hawthorne followed suit. Most of his major works are intrinsically embedded within particular locales and spaces: *The Scarlet Letter* in Salem, *The House of the Seven Gables* at Colonel Pyncheon's house[2] in Salem, *The Blithedale Romance* at George Ripley's utopian Brook Farm, *The Marble Faun* in Rome, and so forth.

While a vast country as the United States of America is marked by a telling plurality, there are certain key aspects and concepts which cut across such pluralities and differences and strike a note of commonality and uniqueness: the American Dream, preoccupation with the Frontier myth, American Exceptionalism, etc. If one were to single out one particular aspect which has been seminal and central to America as a nation, it is possibly space or region. This single aspect of space has proved instrumental in chiselling the dynamics of American history, society, politics, and literature in a considerable way. The very birth of America, or to put it more precisely, the chance discovery of America by Christopher Columbus was propelled by the Renaissance sensibility of exploring new spaces and new worlds. Americans are dislocated Europeans and immigrants from other nations. The history of the genesis of the USA as a nation is premised upon spatial shifts. F.J. Turner's famous article "The Significance of the Frontier in American History" (1893) tried to trace the gradual growth and development of America along the line of the moving frontier, which again is based on the exploration of space along the horizontal axis. The American Dream, as Robert Glen Deamer[3] points out, is 'a cluster of myths which happen, mainly, to be myths of place' (1). And if we have a cursory glimpse at some of the major events of America, like the colonial settlement, the War of Independence, the American Civil War, to adduce only a few examples, we cannot but admit that space is the fulcrum along which the socio-political causes behind these events hinge. No wonder then, space is embedded in the American sensibility as one of its key aspects. Some of the recent theories veer around the concept of space and region in one way or another: colonialism, postcolonialism, imperialism, globalization, nationhood, ecofeminism, ecocriticism, etc.

II

Before we come to the concept of space, it is better to have clarity of the concept of space. As David Harvey[4] puts it:

> 'Space' often elicits modification. Complications sometimes arise from these modifications (which all too frequently get omitted in the telling or the writing) rather than from any inherent complexity in the notion of space itself. When, for example, we write of 'material', 'metaphorical', 'liminal', 'personal', 'social' or 'psychic' space (just to take a few examples) we indicate a variety of contexts that so inflect matters as to render the meaning of space contingent upon the context.

Theories and researches related to space can be examined in several ways. While for the phenomenological critics as Martin Heidegger and Gaston Bachelard, space is inalienably related to one's self, for critics following the Marxist School, as Michel Foucault and Henri Lefebvre, among others, space becomes the site for the contestation of authority, power, and control. Exploring the notions of phenomenology, Heidegger[5] argues that place is dependent on both being and presence:

> Place always opens a region in which it gathers the things in their belonging together. ... Place is not located in a pre-given space, after the manner of physical-technological space.
>
> (6)

Bachelard's concept of space veers around what he calls 'topoanalysis' or 'the systematic psychological study of the sites of our intimate lives,'[6] which is quite different from psychoanalysis. Giving the example of a house Bachelard argues that the space induced by a house can never be neutral and is rather affective. Since it is related to mixed memories of desire, love, compassion, tenderness, sympathy, and so forth, it unconsciously affects us not merely as an objective presence, but as a space that affects and stimulates such emotions, and in the process turns out to be what he calls a 'felicitous space' (xxxv). Bachelard also offers us a distinction between 'exterior space' and 'intimate space' to suggest between physical and psychic space (201).

Another theorist who dwelt on this aspect of space was Foucault. For example, at the beginning of his famous essay "Of Other Spaces" Foucault comments:

> The great obsession of the nineteenth century was, as we know, history ... The present epoch will perhaps above all be the epoch of space. We

> are in the epoch of simultaneity; we are in the epoch of juxtaposition, the epoch of the near and far, of the side-by-side, of the dispersed.
>
> (22)

Interestingly, whenever Foucault talks about the exercise and operation of power relations, he conceives of the same in terms of certain spaces: prison, Panopticon, clinic, and so forth. Foucault conceives of heterotopias as sites 'that have the curious property of being in relation with all the other sites, but in such a way as to suspect, neutralize, or invert the set of relations that they happen to designate, mirror, or reflect' (23). Henri Lefebvre in his interesting research *The Production of Space* tried to use insights from Marxist criticism to suggest how ideology has chiselled the creation of social space. While for Bachelard space is affective, and while for Lefebvre space is ideological, for Michel de Certeau[7] space becomes an apparatus, as it were, for the exercise of hegemonic power.

Denis Donoghue[8] persuasively suggests, 'Place defines a character by confining him to his place. As the story would be another story if its setting were changed.' Wallace Stegner, on the other hand, stresses how places tend to become important in their given literary circumstances. As Stegner[9] claims:

> No place is a place until things that have happened in it are remembered in history, ballads, yarns, legends, or monuments. Fictions serve as well as facts.
>
> (Stegner, 202)

The sense of place, Stegner argues, is at once geographical, economic, and political. It is so vital an aspect as to accommodate other aspects like tradition, taste, sensibility, dialect, and even morality.

Any discussion on space appears incomplete without a reference to Eudora Welty's famous essay "Place in Fiction."[10] Pointing out the indebtedness of a writer to his place, Welty argues that 'place is where he has his roots, place is where he stands; in his experience out of which he writes, it provides the base of reference; in his work, the point of view' (117). Welty claims that fiction is intrinsically embedded in a place that stimulates certain kinds of feelings:

> It is by nature of itself that fiction is all bound up in the local. The internal reason for that is surely that *feelings* are bound up in place.
>
> (118)

The impact of a place on a character, Welty argues, is indisputably true:

> Place, then, has the most delicate control over character too: by confining character, it defines it ... Location pertains to feeling; feeling

> profoundly pertains to place; place in history partakes of feeling, as feeling bout history partakes of place.
>
> (121)

To do justice to the significance of the region in the American context, one must take into account some of the researches carried out in this direction. The two earlier works in this regard are Henry Nash Smith's *Virgin Land* and Leo Marx's *The Machine in the Garden.* Marx examines how the advent of technology has posed a serious threat to the Edenic pastoral ideal of America and how it has shaped the dynamics of American culture substantially. Following in the footsteps of Smith and Marx, Philip Fisher in his research *Hard Facts* examines the construction of place, with its nuanced subtleties, in the works of Harriet Beecher Stowe, Fenimore Cooper, and Theodore Dreiser.

The only two critics, to my poor knowledge, who have tangentially touched upon the aspect of place in Hawthorne, are Gillian Brown[11] and Janice B. Daniel.[12] But while Brown discusses place as an integral concomitant of history and tries to dwell on the vivid delineation of certain places in Hawthorne, the point that is overlooked here is the implication of such places to a better understanding of *The Scarlet Letter*. Similarly, while Daniel's article focuses on the sentient use of Nature in Hawthorne, and while she rightly offers us how Nature has been explored as a living presence in Hawthorne, she fails to point out how place and Nature can be interpreted to examine certain behavioural and attitudinal patterns in some characters in Hawthorne's canon. Given this research gap, my humble submission in this chapter is to substantiate the inextricable correspondence between the physical and the psychological, between the landscape and the mindscape in two representative texts of Hawthorne, "Young Goodman Brown" and *The Scarlet Letter*.

III

Hawthorne was very much conversant with the history[13] and geography of Massachusetts, Boston, and Salem. Hawthorne's keen interest in travelling and exploring new places may be attested by his letters written to his friend H.G. Longfellow. For example, in a letter written to Longfellow, he laments his cloistered place of living and famishes for an excursion:

> I intend, in a week or two, to come out of my owl's nest, and not return to it till late in the summer—employing the interval in making a tour somewhere in New England.[14] In the same letter, he expresses his 'desire to run round the world.'
>
> (Quoted in McIntosh, 297)

Hawthorne's fascination with spaces may be substantiated by his continual shifts across multiple geographical locales in his life: Salem, Raymond, Boston,

Lennox near the Berkshire Mountains; Concord where he purchased Bronson Alcott's house (renamed by him as 'The Wayside'); his stay in Liverpool, England, as the American Consul during 1853–1857; and his extensive tours in France and Italy during 1858–1859. At the age of 12 when their family had shifted to Raymond, Maine, he delighted in the scenic beauty of the landscape around Lake Sebago. But this temporary phase of freedom petered out when he was sent back to Salem for his studies by his mother. In September 1819, he wrote to his sister Louisa, 'I wish I was but in Raymond, and I should be happy,' and thus expressed his grievance against confinement.[15] Hawthorne's penchant for excursions and peregrinations overtook him in later years. As Terence Martin notes, how '[t]he habit of making seasonal trips ... took Hawthorne into the society'; how he enjoyed a trip to New Haven during the fall of 1828 along with his uncle, Samuel Manning; and how in 1831 'he visited New Hampshire and saw the Shaker village in Canterbury' (22).

In an article "Hawthorne's American History," Gillian Brown argues that Hawthorne's sense of history cannot be divorced from his sense of geography and helps enlarge his vision of the world:

> The power of places does not diminish in Hawthorne's revision of the kinship between persons and places. Instead, individuals learn to use the affective power of places as an imaginative engine to take them to different situations, to enlarge and enliven their social connections.
>
> (134)

The same places, when touched upon by imagination, tend to transcend their immediate association with the specific geography and history in which they are situated. Brown argues that rather than confine history to a limited spatio-temporal basis, Hawthorne liberated them, as it were, by conflating imagination with reality, fiction with fact:

> By inventing countries in the inventory of the imaginary, Hawthorne redirects the affective power of place so that it flows from human minds instead of over them. He thereby reconceives the influence of place as an imaginative opportunity.
>
> (134)

Places and Nature were a sentient presence in Hawthorne, and this did not escape the notice of critics and friends. Henry Wadsworth Longfellow, for example, pointed it out in his criticism of the *Twice-Told Tales*:

> But it is one of the high attributes of the poetic mind, to feel a universal sympathy with Nature, both in the material world and in the soul of man.
>
> (Quoted in McIntosh, 329)

Following in the same footsteps, Melville speaks of the sentient Nature in Hawthorne, when he metaphorically comments:

> For no less ripe than ruddy are the apples of the thoughts and fancies in this sweet man of mosses.
>
> (Quoted in McIntosh, 241)

Finally, one may say that the places where Hawthorne had lived conditioned his novels thoroughly. His experience of staying at George Ripley's Brook Farm loomed over his consciousness and acted as 'an available foothold between fiction and reality'[16] in *The Blithedale Romance* (1852). His stay at the Old Manse in Concord inspired his *Mosses from an Old Manse* (1846), while his stay in a house near Berkshire Mountains led to the literary precipitation of *A Wonder-Book for Girls and Boys* (1852) and *Tanglewood Tales for Girls and Boys* (1853). Similarly, his experience of touring in Rome provided a strong impact on him and acted as an impetus for writing *The Marble Faun* (1860). As Hawthorne wrote to his editor James T. Fields that 'no place ever took so stronghold of my being as Rome, nor ever seemed so close to me and so strangely familiar' (quoted in Terence Martin, 161).

IV

Hawthorne's preoccupation with space takes both a symbolic and psychological dimension in his short story "Young Goodman Brown." A descendent of John Hathorne, one of the judges of the Salem Witch Trials of 1692, Hawthorne was so much embarrassed by his descent from this family that he added the 'w' to his title to suit his attitudinal and ideological differences from his strictly puritanical forefathers. The infamous Witch Trials leave a grim impact on "Young Goodman Brown" in which Goodman Brown sets out for the forest on an evil errand, refusing the request of her faithful wife Faith. In the forest rendezvous, he meets an old man with a staff resembling a black snake. While the old man acts like the devil, his serpentine staff symbolizes temptation by alluding to Satan's temptation of Even under the disguise of a serpent. In no tale by Hawthorne does one find the rendering of evil as vivid as in this one. Goodman Brown's assignation with the old man through the deep labyrinthine forest may be compared with Marlow's journey through the gloom of the Congo in Conrad's *Heart of Darkness*, for this journey metaphorically corresponds to the psychological exploration into the deepest recesses of the human unconscious in pursuit of the nature of evil. Furthermore, it also bears resemblance to Dante's journey through the different circles of hell with the senior poet Virgil in "Inferno" of *Divine Comedy*. As Melville in his essay "Hawthorne and His Mosses" so pertinently points out, "Young Goodman Brown" is 'deep as Dante' and that it is 'a strong positive illustration of the blackness in Hawthorne' manifested so vividly (quoted in McIntosh, 348).

The very description of the forest, 'darkened by the gloomiest of trees,' is shot with mystery and rendered in demonic terms (66). Goodman Brown apprehends that there might be a 'devilish Indian' lurking behind every tree.

His initial reluctance to go with the man further into the forest on the pretext that his father had never been on such errands into the woods and that he hails from 'a race of honest men and good Christians,' along with his desire to return home, presents the very forest as an embodiment of its antithesis of home (67). Given the spatial preoccupations of Hawthorne, we may say that Goodman Brown's home with Faith stands for innocence, goodness, and fidelity, while the forest exemplifies experience, evil, and infidelity spurred by temptation. Goodman Brown's journey through the forest symbolizes a journey from innocence to experience. The deeper he delves into the forest, the more is he divested of his hitherto innocence. He is shocked to find that the pious Goody Cloyse from whom he had learnt catechism in the forest. He is further aghast to find, from his clandestine position, that she is a witch and that his companion is the devil. The presence of the good old Deacon Gookin in this forest leads Goodman Brown to further disenchantment since it becomes clear to him that all of them have assembled in the forest to join the devil's bacchanalia. Thus, on the one hand, the forest exemplifies the very embodiment of evil with all its impenetrability, paralleling the dense impenetrability of the forest; on the other, the forest also stands for revelation and disillusionment. It is in the forest that Goodman Brown fathoms the acuity of evil in its absolute form. It is in the forest again that he realizes the true nature of some of his acquaintances whom he had hitherto held from an innocent perspective. As Terence Martin puts it:

> He (Goodman Brown) has been exposed to the mightiest vision of evil that Hawthorne ever imagined, and it sears and blights the very roots of its being.
>
> (89)

The indistinct 'voice of a young woman, uttering lamentations, yet with an uncertain sorrow, and entreating for some favour' adds to the gruesomeness of evil and smacks of some insidious, sinister omen to Goodman Brown (71). His agony and desperation knew no bound when he identified the voice to be that of his faithful wife Faith. Bewildered, he began to acquiesce in the inevitability of evil as an irrevocable human predicament: 'There is no good on earth, and sin is but a name. Come, devil, for to thee is this world given' (71).

Disillusioned and disenchanted, Goodman Brown behaved in a frenzied manner, bordering on insanity. The forest reveals before him the deepest form of evil, a revelation that crippled him psychologically to bewilderment and disintegration. Hawthorne wonderfully describes his frenzied bewildered behaviour in spatial terms by comparing it with the wilderness of the forest:

> The road grew wilder and drearier and more faintly traced, and vanished at length, leaving him in the heart of the dark wilderness ... The

> whole forest was peopled with frightful sounds ... as if all Nature were laughing him to scorn.
>
> (71)

In keeping with his penchant for darkness, mystery, and indistinctness, Hawthorne suggests that Goodman Brown might have fallen asleep in the forest and might have had this horrible dream of the meeting of witches and the devil. The next morning after his return to Salem village, his bewilderment did not desert him. Neither could he meet Goody Cloyse and the old Deacon Gookin naturally, nor could he encounter his faithful wife Faith with a greeting. Thus, his vision of evil in the wilderness of the forest makes him an unknown man groping through the wilderness of his heart of darkness.

Broadly speaking, "Young Goodman Brown" veers around three prominent spaces: Goodman Brown's home, the forest, and the Salem village. The home and the forest are antithetically poised, and they bring out two opposite sets of values belonging to two opposite paradigms. While the home stands for stability, fidelity, goodness, innocence, and established positive values, the forest amply exemplifies the essence of evil, loss of innocence, wilderness of human psyche, bordering on the Freudian Id, and precariousness. The forest also represents man's temptation, his irredeemable fall of man, and his precarious predicament. Broadly speaking, these three spaces correspond to the three categories of the human unconscious referred to by Freud: the forest corresponds to the Id, for Goodman Brown is irresistibly drawn towards it, and is actuated by a pleasure principle; the Salem village corresponds to the Ego, since it refers to the awareness of the identity of the characters as they are supposed to be, and it is dominated by reality principle; and the home of Goodman Brown corresponds to the Superego, since it is the abode of goodness and innocence and is governed by morality principle.

V

The introductory chapter of *The Scarlet Letter*, "The Custom House," contains the seed of the entire novel and its multiple aspects. Based on Hawthorne's job as a surveyor at the Salem Custom House during 1846–1849, it succinctly captures the social picture of contemporary times, along with occasional snapshots of his personal life. The vital role of a place in shaping one's sensibility and attitude is borne out by the following comment of the narrator:

> This long connection of a family with one spot, as its place of birth and burial, creates a kindred between the human being and the locality, quite independent of any charm in the scenario or moral circumstances that surround him. It is not love but instinct.
>
> (9)

Rather than present a place in its neutral, inert, objective way, Hawthorne describes a place being capable of evoking certain emotions:

> It is no matter that the place is joyless for him … The spell survives, and just as powerfully as if the natal spot were an earthly paradise.
>
> (9)

Similarly, the correspondence between physical space and human behaviour is conceived of in spatial terms with an organic metaphor:

> Human nature will not flourish, any more than a potato, if it be planted and replanted for too long a series of generations in the same worn-out soil.
>
> (10).

Some of the prominent places delineated in the novel are the prison house, the Market Place, Governor Bellingham's Hall, and the forest, and each of them has a specific unique nature of its own. For example, the ponderous door of the prison house was 'heavily timbered with oak and iron spikes,' the iron spikes being symptomatic of its intrinsic cruel nature and its harshness (43). But what strikes our attention is the existence of the wild rosebush which, in Hawthorne's words, may symbolize 'some sweet moral blossom that may be found along the track, or relieve the darkening close of a tale of human frailty and sorrow' (44). W.H. New in his article "Beyond Nationalism: On Regionalism" argues:

> [d]efining the significance that particular spaces have for us, we parenthetically define or at least hint at our preconceptions about the significance of statements that will emerge from those given spaces, locales, regions.
>
> (14)

What New suggests seems to be all the more applicable to Hawthorne whose representation of a place throbs with symbolic significance. The marketplace, as the name suggests, is an open space susceptible to the discussions and gossips of common people. One may recall how in Chapter 2 ("The Market Place") many people thronged in the marketplace and how they were passing their opinions and judgements on Hester's punishment, prior to her entry into the marketplace from the prison door. One may be tempted to add, the marketplace, because of its open access to the multitude, becomes a space where the voices of the common people function very much like the collective voice of the Chorus in a Greek tragedy. Like a Chorus which relates the past, comments on the present, and anticipates the future without directly getting involved in the main action, from the snatches of the gossips brewing around the marketplace one come to know of the immediate past of Hester's adultery, of her present ensuing trial, and her impending punishment awaiting her in the future. We are also referred to 'the Reverend Master Dimmesdale, her godly pastor' who 'takes it grievously that such a scandal should have come upon his congregation' (47). We are exposed to a wide variety of conjectures as to the nature and intensity of Hester's punishment well ahead of the cruel judgement is meted out to her.

If the marketplace is marked by openness and access, the otherwise inaccessible hall of Governor Bellingham is couched in metaphors of shabbiness, bleakness, decay, and remoteness:

> This was a large wooden house, built in a fashion of which there are specimens still extant in the streets of our elder towns now moss-grown, crumbling to decay, and melancholy at heart with the many sorrowful or joyful occurrences, remembered or forgotten, that have happened and passed away within their dusky chambers.
>
> (96)

One cannot but notice an asphyxiating sombreness in the fact that the portraits of Governor Bellingham's ancestors are 'characterized by sternness and severity,' bespeaks their cruelty, and rigidity (98).

Even when Hawthorne describes the dejected minister, Arthur Dimmesdale, he does so in spatial terms. Dimmesdale's house is, as it were, analogous to his self, the innermost chamber of which is his heart. Roger Chillingworth, by exercising his acute, perceptive power, rightly suspects Dimmesdale to be the partner of Hester's adultery and zeroes in on his revengeful onslaughts on him. Claiming himself to be a physician invaluable to the minister, Chillingworth managed the friends of Dimmesdale to make 'an arrangement by which the two were lodged in the same house; so that every ebb and flow of the minister's life-tide might pass under the eye of his anxious and attached physician' (118). Through his watchful penetrative gaze, Chillingworth wanted to 'burrow into the clergyman's intimacy and plot against his soul' (121). Chillingworth's frantic and obsessive search to find out the hidden secret from the innermost recess of the minister's mindscape has been conceived of in terms of the metaphor of the excavation of a landscape in search of wealth:

> He(Chillingworth) now dug into the poor clergyman's heart like a miner searching for gold; or, rather, like a sexton delving into a grave, possibly in quest of a jewel that had been buried on the dead man's bosom.
>
> (122)

One does not fail to notice that even the title of Chapter 11, "The Interior of a Heart," is conceived of in spatial terms and reinforces the suggestion of finding out the psychological only through an exploration of the physical, of the mindscape through an excavation of the landscape. Besides, the very titles of a few other chapters – the introductory chapter, "The Custom House," Chapter 1 ("The Prison House"), Chapter 2 ("The Market Place"), Chapter 7 ("The Governor's Hall"), Chapter 16 ("A Forest Walk"), and

Chapter 19 ("The Child at the Brookside") – make conspicuous mention of certain spaces, while all the major chapters of this novel are located in certain spaces which convey deeper significance than being just physical spaces.

The Scarlet Letter offers an excellent scope of deep psychological analysis of its major characters. It is the ambivalence of his characters that demands a deeper psychological probe into them. As Terence Martin cogently points out that Hawthorne's 'art, at its best, thrived on ambivalence' (18). 'The special quality of Hawthorne's achievement in *The Scarlet Letter*,' observes Martin, 'inheres in the essential duality or ambivalence of his fictional world and the components of that world' (119). All the three characters forming the adulterous triangle – the adulterous heroine Hester Prynne, her ex-husband Roger Chillingworth, and her pastor and lover Arthur Dimmesdale – may be examined as having split personalities, and each of them is sharply divided into his/her private self and public self. Under the seeming serenity and placidity of Hester's essential feminine nature lurks her essential untamed adulterous self. Similarly, her quiet demeanour notwithstanding, one does not fail to notice her anarchic nature which defies blurting out the name of her adulterous partner under duress and constant provocations of the puritanical ministers. As the narrator puts it:

> 'Never!' replied Hester Prynne, looking, not at Mr. Wilson, but into the deep and troubled eyes of the younger clergyman. 'It is too deeply branded. Ye cannot take it off. And would that I might endure his agony, as well as mine!'
>
> (63)

Mr Dimmesdale's comment is more of a commendation of her character:

> Wondrous strength and generosity of a woman's heart!
>
> (64)

Roger Chillingworth is also a split personality in terms of the stark dichotomy between his public self and private self. His public self reveals him to be an old innocent and innocuous physician, one who is a bookworm and is absorbed in his study. And yet, his is the study of revenge. Thus, his private self betrays his true evil and insidious nature. Consumed by the fire of revenge, he embodies the sin of intellect. Under the veneer of his public self of a friend and physician of Dimmesdale, he turns out to be the minister's avowed enemy, one who sadistically runs into rhapsodies over the anguish he inflicts on the minister. As the narrator puts the equation between the two:

> To sum up the matter, it grew to be a widely diffused opinion that the Reverend Arthur Dimmesdale, like many personages of special sanctity,

> in all ages of the Christian world, was haunted either by Satan himself, or Satan's emissary, in the guise of old Roger Chillingworth.
>
> (121)

But the greatest of these duplicitous characters is indisputably Arthur Dimmesdale. While his public self of a minister punishes Hester on the first scaffold scene, his inner private self is thoroughly lacerated by the lashes of his guilt. While the outer public self of Dimmesdale could not embrace the punishment with Hester by confessing his adultery, the inner private self of him famishes for confession.

My proposition in this chapter is to suggest that the marketplace stands for the public self of both Hester Prynne and Arthur Dimmesdale. Just consider the first scaffold scene during Hester's public trial. Unable to come out from the veneer of a puritanical minister, Dimmesdale – also a gifted speaker marked by the facility and felicity of his speech – concocts a double-edged language to interrogate Hester:

> I charge thee to speak out the name of thy fellow-sinner and fellow-sufferer! Be not silent from any mistaken pity and tenderness for him; for, believe me, Hester, though he were to step down from a high place, and stand there on thy pedestal of shame, yet better were it so, than to hide a guilty heart through life.
>
> (62)

Any perceptive reader cannot but notice that none but the sufferer alone knows whether he suffers or not, and none but the person concerned can know whether he has a guilty heart or not. Thus, through this double-edged interrogation, he has two meanings for the multitude and Hester. Furthermore, so far as Dimmesdale is concerned, we may add more precisely that it is the marketplace during the day that helps him maintain his public self. The same marketplace during the night helps him give vent to his inner and private self. In Chapter 12 ("The Minister's Vigil"), we encounter the second scaffold scene, here eclipsed in the darkness of night. When the entire town was lulled into delicious slumber, the minister surreptitiously ascends the steps of the same scaffold where he had punished Hester seven years back, an attempt that acts as a sort of prelude to his final public confession. As the narrator describes the scene:

> He (Dimmesdale) had been driven hither by the impulse of that Remorse which dogged him everywhere, and whose own sister and closely linked companion was that Cowardice which invariably drew him back with her tremulous gripe, just when the other impulse had hurried him to the verge of a disclosure.
>
> (141)

Ironically, Dimmesdale's private self finds an outlet only under the shroud of darkness, while the broad daylight serves to eclipse his real inner private self.

If the marketplace stands for the public selves of Dimmesdale and Hester, the forest stands for their private selves. It is in the forest assignation that both Dimmesdale and Hester unleash their hearts. In Chapter 17 ("The Pastor and His Parishioner"), in the forest, both Hester and Dimmesdale give free vent to the innermost recess of their private selves. We find vivid glimpses of the minister's heart rocked and racked by despair, battered and buffeted by guilt, and tossed and tormented by remorse. Almost in a confessional manner, the minister says:

> As concerns the good I may appear to do, I have no faith in it. It must needs be a delusion. What can a ruined soul like mine effect towards the redemption of other souls?—or a polluted soul, towards their purification?
>
> (184)

His excruciating pain induced by the lashes of his guilt finds expression only in the forest:

> Happy are you, Hester, that wear the scarlet letter openly upon your bosom! Mine burns in secret!
>
> (185)

It is in the forest again that Hester 'threw her arm around him, and pressed his head against her bosom; little caring though his cheek rested on the scarlet letter' (187). Thus, Hester unleashes her passion for her pastor-turned-paramour only in the forest. It is in the forest again that they ratify their previous adultery and justify it as a sacred act:

> What we did had a consecration of its own. We felt it so! We said so to each other!
>
> (188)

The forest scenes conduce to the unravelling of one's inner private self to the fullest possible degree. It is in the forest that Hester looks forward to having a domestic life with Dimmesdale and projects the same before Pearl:

> We will have a home and fireside of our own; thou shalt sit upon his knee; and he will teach thee many things, and love thee dearly. Thou wilt love him; wilt thou not?
>
> (205)

She dreams of a better future in her domestic life with Dimmesdale and tries to invigorate his drooped and dejected self:

> Begin all anew! … . The future is yet full of trial and success. There is happiness to be enjoyed! There is good to be done!
>
> (191)

One may be tempted to add that this personal dream of Hester is but a small constituent of what constitutes her American Dream.

If the Forest of Arden in Shakespeare's *As You Like It* is antithetically poised against the sneers and snares of the court, the forest for Hester Prynne stands for freedom and liberty. Apart from being a space for the expression of her private self, the forest epitomizes freedom for Hester. Just as the public self is antithetical to the private self, so also the marketplace is antithetical to the forest. It is in the forest that Hester tries to find out relief from the excruciating torments induced by the scarlet letter by unfastening and throwing away the scarlet letter itself. This was followed by letting her confined hair fall to its full length, an act that symbolizes liberty and relief. It was an attempt to assert her freedom from her social ostracism and to reclaim her love for Dimmesdale. As the narrator puts it so wonderfully:

> The stigma gone, Hester heaved a long, deep sigh, in which the burden of shame and anguish departed from her spirit. O *exquisite relief! She had not known the weight until she felt the freedom!* By another impulse, she took off the formal cap that confined her hair; and down it fell upon her shoulders, dark and rich, with at once a shadow and a light in its abundance, and imparting the charm of softness to her features.
>
> (195, emphasis added)

She felt a replenishment of her love and rejuvenation of her instinctive faculties only in the forest:

> Had the forest kept its gloom, it would have been bright in Hester's eyes, and bright in Arthur Dimmesdale's!
>
> (196)

The same holds for Dimmesdale who finds a fulfilment of love only in the forest which unearths his private inner self. His filial affections are stimulated and realized in the forest. For the first time in these seven years, he touches Pearl and plants a kiss on her forehead.

But the inherent duality of the minister does not elude the keen ken of Pearl who points out his duplicitous nature to her mother:

> 'What a strange, sad man is he!' said the child, as if speaking partly to herself. '*In the dark night-time he calls us to him, and holds thy hand and mine, as we stood with him with the scaffold yonder!* And in the deep forest where only the old trees can hear and the strip of the sky

> see it, he talks with thee, sitting on a heap of moss! And he kisses my forehead, too, so that the little brook would hardly wash it off! *But here, in the sunny day, and among all the people, he knows us not*; nor must we know him! A strange sad man is he, with his hand always over his heart!'
>
> (221, emphasis added)

Pearl acts as a touchstone, as it were, to fathom the real ebb and tide flowing within a character's psychological recesses. Ironically enough, it is not the sunshine of the marketplace but the gloom of the forest that unravels layers of truth and lays bare the true private selves of both Hester and Dimmesdale. The forest, along with its privacy, allows Hester to disclose the fact that Chillingworth was her husband.

It is Chapter 23 ("The Revelation of the Scarlet Letter") where we also have the final scaffold scene. If the first scaffold scene in Chapter 2 ("The Market Place") corresponds to Dimmesdale's public self as a puritanical minister, and if the second scaffold scene in Chapter 12 ("The Minister's Vigil") corresponds to his private self-ascending the scaffold at midnight in an attempt of confession, it is this final scaffold scene where Dimmesdale's public and private selves fuse and coalesce into complete unity, as it were. The hitherto dichotomy and invisible dividing line between his two otherwise incompatible selves gets dissolved, as it were, and Dimmesdale is elevated to a higher pedestal of embracing and confessing the truth in a triumphant moment on the scaffold along with his death. Hawthorne's use of the scaffold as space thus becomes amply significant. The three scaffold scenes correspond to three different attitudes and phases of Dimmesdale: his hypocrisy and self-deception in the first, the pricking of his conscience and his struggle towards confessing the truth in the second, and his overcoming the barrier of deception and hypocrisy and his triumphant embracing of truth in the third. Finally, after the death of Hester, even the description of her graveyard is rendered in spatial terms:

> And after many, many years, a new grave was delved, near an old and sunken one, in that burial-ground beside which king's Chapel has since been built. Yet was near that old and sunken grave, *yet with a space between*, as if the dust of the two sleepers had no right to mingle.
>
> (255, emphasis added)

This puritanical rigidity finds an ingenious spatial representation through the cachet of Hawthorne's genius.

Conclusion

'Feelings are bound up in place,' claims Eudora Welty and certain spaces in *The Scarlet Letter* are evocative of certain feelings and attitudes (123).

In a similar vein, but an almost aphoristic way Louis D. Rubin, Jr.[17] claims that 'all places in fiction are potentially sensory' (14). One cannot but admit the impact of region and place – or what Robert Maria Dainotto calls 'a genealogy of the literature of place and region'[18] in American literature. During Hawthorne's time and particularly during the seventeenth century in which *The Scarlet Letter* is set, America was dominated by wilderness. In both "Young Goodman Brown" and *The Scarlet Letter* the wilderness is represented by the forest which stands for both fright and freedom. While Goodman Brown and Dimmesdale shrink from the fright, Hester embraces the freedom. Thus, space becomes so pervasive and plays so vital a role in "Young Goodman Brown" and *The Scarlet Letter* that its presence is felt throughout these texts, running through the whole gamut of varied human experiences: love, adultery, punishment, suffering, conflict, temptation, fall, struggle, and death.

Notes

1 D.H. Lawrence who in his *Studies in Classic American Literature* lays bare its duplicitous undermeaning. Under the ostensible veneer of its placidity, Lawrence argues, lurks the 'inner diabolism and symbolic meaning' of its narrative (78). Both R.W.B. Lewis and Leslie Fiedler prefer to read it as a re-enactment of an Edenic myth and the fall of man in a new Eden. Following in the same line, Leslie Fiedler in *Love and Death in the American Novel* argues that '*The Scarlet Letter* is the most utopian of American books: not the Paradise Regained it seems at first, but only an Eden Revisited' (232). Among recent critics of Hawthorne, Richard H. Brodhead in *The School of Hawthorne* makes an assiduous attempt to situate Hawthorne in the American tradition and tries to redefine the tradition. In his thorough and prolific research, Sacvan Bercovitch in *The Office* of The Scarlet Letter explores the patterns of ambiguity, paradox, and irony in *The Scarlet Letter*. In *Secrets and Sympathy: Forms of Disclosure in Hawthorne's Novels* Gordon Hunter examines Hawthorne's strategies of disclosure vis-à-vis his intrinsic art of concealment. Kenneth M. Harris's research *Hypocrisy and Self-Deception in Hawthorne's Fiction* dwells on the dual aspects of self-deception in Hawthorne and argues, 'In *The Scarlet Letter*, as elsewhere in his fiction, and here as much as anywhere, we are in a world pervaded by hypocrisy and self-deception' (46).

2 The house is still there and has been turned into a sort of museum. It is close to Hawthorne's own house in Salem.

3 Robert Glen Deamer, *The Importance of Place in the American Literature of Hawthorne, Thoreau, Crane, Adams, and Faulkner: American Writers, American Culture, and the American Dream* (Edwin Mellen Press, 1990).

4 David Harvey, 'Space as a key word.' In *Spaces of Global Capitalism: Towards a Theory of Uneven Geographical Development* (Verso, 2006), 119.

5 Martin Heidegger, "Art and Space". Trans. Charles Seibert. *Man and World* 1(1973): 3–8.

6 Bachelard, *The Poetics of Space*, trans. Maria Jolas (Beacon, 1994), 8.

7 Michel de Certeau, *The Practice of Everyday Life*. Trans. Steven Rendall (University of California Press, 1984).

8 Denis Donoghue, "Eudora Welty's Sense of Place", in *Place in American Fiction: Excursions and Explorations*, eds. H.L. Weatherby, and George Core (University of Missouri Press, 2004), 142.

9 Wallace Stegner, "The Sense of Place", in *Where the Bluebird Sings to the Lemonade Springs: Living and Writing in the West* (Random House, 1992), 202.
10 Eudora Welty, *The Eye of the Story: Selected Essays and Reviews* (Random House, 1978), 116–133.
11 Gillian Brown, "Hawthorne's American history", in Richard H. Millington, ed. *The Cambridge Companion to Nathaniel Hawthorne* (Cambridge University Press, 2004), 121–142.
12 Janice B. Daniel, '"Apples of the Thoughts and Fancies": *Nature as Narrator in* The Scarlet Letter', in Harold Bloom, ed. *Nathaniel Hawthorne's* The Scarlet Letter (Viva Books, 2010), 95–108.
13 See Charles Ryskamp "The New England Sources of *The Scarlet Letter*", in *Nathaniel Hawthorne:* The Scarlet Letter *and Other Writings*, ed. Leland S. Person (W.W. Norton & Company, 2005), 291–303. Ryskamp tells us how Hawthorne was conversant with Bancroft's *History of the United States*, Dr Caleb H. Snow's *History of Boston*, Hutchinson's *History of Massachusetts*, Felt's *Annals of Salem*, among other books (292). One may also consult Charles Boewe and Murray G. Murphy, "Hester Prynne in History" From *American Literature* 32 (1960): 202–204.
14 Quoted in *Nathaniel Hawthorne's Tales*, ed. James McIntosh (W.W. Norton & Company, 1987), 298.
15 Quoted in Terence Martin, *Nathaniel Hawthorne* (Twayne Publishers, 1965), 17.
16 Terence Martin, 39.
17 Louis D. Rubins, Jr. "Thoughts on Fictional Places" in *Place in American Fiction: Excursions and Explorations*, eds. H. L. Weatherby, and George Core (University of Missouri Press,2004), 13–21.
18 Robert Maria Dainotto, *Place in Literature: Regions, Cultures, Communities* (Cornell University Press, 2000), 30.

Works Cited

In this paper I have used *Nathaniel Hawthorne's Tales* edited by James McIntosh as the primary text for "Young Goodman Brown," and I have followed the Watermill Press version of *The Scarlet Letter* (1983) as my primary source.

Bachelard, Gaston. *The Poetics of Space*. Trans. Maria Jolas. Beacon, 1994.

Bercovitch, Scavan. *The Office of* The Scarlet Letter. The Johns Hopkins University Press, 1991.

Boewe, Charles and Murray G. Murphy. "Hester Prynne in History." *American Literature*. 32 (1960): 202–204.

Bloom, Harold ed. *Nathaniel Hawthorne's The Scarlet Letter*. Viva Books, 2010.

Brodhead, Richard H. *The School of Hawthorne*. Oxford University Press, 1986.

Certeau, Michel de. *The Practice of Everyday Life*. Trans. Steven Rendall. University of California Press, 1984.

Dainotto, Robert Maria. *Place in Literature: Regions, Cultures, Communities*. Cornell University Press, 2000.

Deamer, Robert Glen. *The Importance of Place in the American Literature of Hawthorne, Thoreau, Crane, Adams, and Faulkner: American Writers, American Culture, and the American Dream*. Edwin Mellen Press, 1990.

Fiedler, Leslie. *Love and Death in the American Novel*. Criterion Books Inc., 1960.

Fisher, Philip. *Hard Facts: Setting and Form in the American Novel*. Oxford University Press, 1985.

Foucault, Michel. 'Of Other Spaces.' *Diacritics*. Vol. 16. 1 (Spring 1986): 22–27.
Harris, Kenneth Marc. *Hypocrisy and Self-Deception in Hawthorne's Fiction*. University Press of Virginia, 1988.
Harvey, David. "Space as a Key Word." *Spaces of Global Capitalism: Towards a Theory of Uneven Geographical Development*. Verso, 2006.
Hawthorne, Nathaniel. *The Scarlet Letter*. Watermill Press, 1983.
Heidegger, Martin. "Art and Space". Trans. Charles Seibert. *Man and World* 1(1973): 3–8.
Hunter, Gordon. *Secrets and Sympathy: Forms of Disclosure in Hawthorne's Novels*. The University of Georgia Press, 1988.
Jackson, Robert. *Seeking the Regions in American Literature and Culture: Modernity, Dissidence, Innovation*. Louisiana State University Press, 2005.
Lawrence, D. H. *Studies in Classic American Literature*. Heinemann, 1924.
Lefebvre, Henri. *The Production of Space*. Trans. Donald Nicholson-Smith. Blackwell, 2007.
Lewis, R.W.B. *The American Adam: Innocence, Tragedy, and Tradition in the Nineteenth Century*. University of Chicago Press, 1955.
Martin, Terence. *Nathaniel Hawthorne*. Twayne Publishers, Inc, 1965.
Marx, Leo. *The Machine in the Garden: Technology and the Pastoral Ideal in America*. Oxford University Press, 1964.
McIntosh, James. ed. *Nathaniel Hawthorne's Tales*. W.W. Norton &Company, 1987.
Millington, Richard H. ed. *The Cambridge Companion to Nathaniel Hawthorne*. Cambridge University Press, 2004.
New, W.H. "Beyond Nationalism: On Regionalism". *World Literature Written in English* 25.1(1984): 12–17.
Person, Leland S. ed. *Nathaniel Hawthorne: The Scarlet Letter and Other Writings*. W.W. Norton & Company, 2005.
Smith, Henry Nash. *Virgin Land: The American West as Symbol and Myth*. Harvard University Press, 1950.
Stegner, Wallace. *Where the Bluebird Sings to the Lemonade Springs: Living and Writing in the West*. Random House, 1992.
Weatherby, H.L. and George Core. eds. *Place in American Fiction: Excursions and Explorations*. University of Missouri Press, 2004.
Welty, Eudora. *The Eye of the Story: Selected Essays and Reviews*. Random House, 1978.

11 Absolute Trust in the Goodness of the Earth

An Ecocritical/Ecofeminist Reading of Alice Walker's *The Color Purple*

Anindita Chatterjee

I

Greg Garrard in the introduction to *Ecocriticism* goes on to observe, 'It is generally agreed that modern environmentalism begins with "A Fable for Tomorrow," in Rachel Carson's *Silent Spring* (1962). Carson's fairy tale begins as 'There was once a town in the heart of America where all life seemed to live in harmony' and to depict this concord the author draws upon the ancient tradition of the pastoral and goes on to paint a picture of 'prosperous farms,' 'green fields,' foxes barking in the hills, silent deer, ferns and wildflowers, 'countless birds' and trout lying in clear, cold streams, all delighted in these 'who pass through the town' (Garrard, 2004, 1). Concentrating on the descriptions of natural beauty and emphasizing the 'harmony' of humanity and nature that 'once' existed, the fable introduces us to an idyllic world of pristine bliss and eternal changelessness. However, this primeval image of pastoral peace rapidly gives way to catastrophic destruction due to the impact of urbanization and progress of science. The rural idyll is brutally shattered by some agency of change which is identified either as a malady or an apocalyptic spell.

> It is not surprising that ecocriticism first emerged in the United States, because Americans have been obsessed with the landscapes of the 'New World' since European exploration of the continent began.
>
> (Gersdof, 2006, 26)

Ecocriticism explores the relationship between literature and the physical environment, which necessitates looking at literary studies through a green lens. Ecocriticism deals with the study of the relationship between literature and the physical environment. Whereas feminist criticism examines language and literature from a gender-conscious perspective and Marxist criticism engenders an awareness of modes of production and economic class to its interpretation of texts, ecocriticism is based on an earth-centred approach to literary studies. As a critical enterprise, ecocriticism is deeply entrenched in the environmentalist revisions of US-American nature writing

DOI: 10.4324/9781003383970-14

and nineteenth-century Transcendentalism (with a particular emphasis on Emerson, Thoreau, and Fuller) and of the British institution of late eighteenth-century Romanticism (most conspicuously represented by the figure of William Wordsworth).

> Despite the broad scope of inquiry and disparate levels of sophistication, all ecological criticism shares the fundamental premise that human culture is connected to the physical world, affecting it and affected by it. Ecocriticism takes as its subject the interconnections between nature and culture, specifically the cultural artifacts of language and literature. As a critical stance, it has one foot in literature and the other on land; as a theoretical discourse, it negotiates between the human and the nonhuman. Ecocriticism expands the notion of 'the world' to include the entire ecosphere.
>
> (Glotfelty, 1995, xix)

The term 'ecofeminism' was first used by Françoise d'Eaubonne in her book, *Le Féminisme ou la Mort* (*Feminism or Death*), which was published in 1974. It is a product of the conjunction of two words – ecology and feminism. The term denotes neither a single movement nor philosophy. It is a loosely knit philosophical concept that examines and critiques the mutual devaluation and oppression of women and nature. Various perspectives of ecofeminism mutually coexist. Ecofeminism is multifaceted and multilocated, challenging established structures rather than individuals.

> Although the word 'ecofeminism' is a singular noun, it refers not to a monolithic or homogeneous ecofeminism but to a plurality of ecofeminisms. It refers to a multiplicity of theories, methods, and practices for interpreting and transforming relationships between the bodies that share this same planetary home.
>
> (Vakoch and Mickey, 2018, ix)

By confronting systems of conventional notions of patriarchy, ecofeminism attempts to broaden the scope of the cultural critique and incorporates seemingly disparate but, according to ecofeminism, radically connected elements. However, linguistically joining these two words was also subjected to substantial debate and criticism. In the opinion of some critics being essentialist, the term denotes a gender divide. In Western patriarchal society, conventionally, women are treated as inferior to men, and similarly 'nature' is treated as inferior to 'culture'; women's physiological connection with birth, nurturing, and child care resulted in the close association of women or the feminine with nature. It is often argued that many women in this camp ended up unintentionally perpetuating patterns of division and exclusion that feminism aims to contest. Another criticism of this philosophy is that it holds women alone as socially responsible for environmental protection and care. This controversial notion created a heated debate within feminist literature during the 1980s and 1990s.

Karen J. Warren in her study *Feminism and Ecology: Making Connections* claimed that just as there is no one genre of feminism, there is not one uniform idea of ecofeminism either.

> Ecological feminism is an umbrella term which captures a variety of multicultural perspectives on the nature of the connections within social systems of domination between those humans in subdominant or subordinate positions, particularly women and the domination of non-human nature ... Ecological feminism is the name given to a variety of positions that have their roots in different feminist perspectives (e.g. liberal, traditional Marxist, radical, socialist, Black and Third World); for they reflect different understandings of the nature of and solution to the pressing environmental problems.
>
> (Warren, 1987, 4)

Thus, it is helpful to think of the term ecofeminism as a critique that draws parallels between the subjugation of women and oppression of nature. It refers to a variety of woman–nature connections – historical, empirical, conceptual, religious, literary, political, ethical, epistemological connections on how one treats women and the earth, 'ecofeminist analyses of the twin domination of women and nature include considerations of the domination of people of color, children and the underclass' (Warren, 1994, 1).

> Walker herself highlights throughout her poetry, fiction and polemical writings such connections between social systems of domination of people of color, particularly women, and non-human nature. Her collection of poems *Horses Make a Landscape Look More Beautiful* was inspired by a move from Brooklyn to northern California which was as Walker records enabled her for the first time to admit and express her grief over the ongoing assassination of the earth, even as she accepted all the parts, good and bad of her own heritage.
>
> (Greenwood, 2000, 165)

In *Ecofeminism* (1993), author Vandana Shiva challenges modern science and its acceptance as a universal and value-free system. Instead, the book attempts to understand the dominant streams of modern science, the reductionist or mechanical paradigm, as a particular projection of Western men, originating during the scientific revolution. A common premise of ecofeminist literature is that patriarchal structures justify their dominance through hierarchical dualisms and establish binaries between heaven/earth, mind/body, male/female, human/animal, spirit/matter, culture/nature, and white/non-white. Oppression and subjugation are reinforced by assuming truth in these binaries and instilling them as 'sacred' through religious and scientific constructs. Ecofeminists draw a parallel between the oppression of the earth and exploitation of women. The relationship between man and nature in the Western tradition is distinctly anthropocentric whereby it is believed that the entire nature exists for human benefit alone. Ecofeminism as a theory has evolved from numerous arenas of feminist activism, amity movements, labour movements, anti-nuclear, environmental, animal liberation movements, etc.

> Drawing on the insights of ecology, feminism, and socialism, ecofeminism's basic premise is that the ideology which authorizes oppressions such as those based on race, class, gender, sexuality, physical abilities, and species is the same ideology which sanctions the oppression of nature. Ecofeminism calls for an end to all oppressions.
>
> (Gaard, 1993, 1)

Despite the important differences among them and the feminisms from which they gain their inspiration, ecofeminists like Rosemary Ruether, Ivone Gebara, Vandana Shiva, Susan Griffin, Alice Walker, Starhawk, Sallie McFague, Luisah Teish, Sun Ai Lee Park, Paula Gunn Allen, Greta Gaard, Karen Warren, and Andy Smith have tried to trace the connection between domination of nature and domination of women, an understanding of which is crucial to feminism, environmentalism, and environmental philosophy. The main thrust of ecofeminism is to study woman–nature connections and to expose, criticize, and dismantle those aspects which prove to be harmful to women and nature which include the dehumanizing impact of science, culture, mechanization, and urbanization. Both ecologists and feminists distinguish between the privileged and the oppressed sections of the society, where the privileged constitutes the upper- or middle-class, human, technologically empowered and scientifically and industrially 'developed,' male, and the oppressed constitutes the poor, the marginalized other species, elements of nature, and female gender.

> Ecofeminist writers sometimes claim a global affinity between women as a gender and the movement to save the environment. Women all over the world are demonstrating an intense personal concern over human and natural exploitation. Certain1y, ecofeminism sees women as likely supporters of environmental politics. Since women are life givers, hence their connection to Mother Earth is important.
>
> (Leahy, 2003, 107)

Environmental science historian Carolyn Merchant in her highly influential book *The Death of Nature: Women, Ecology, and the Scientific Revolution* (1990) claimed that before the seventeenth century, nature was conceived on an organic model as a benevolent female and nurturing mother, but after the scientific revolution nature was conceived on a mechanistic model as a mere machine, inert and dead. On both models, nature was female. Merchant goes on to argue how the movement from the organic to the mechanistic model justified the exploitation of the (female) earth. The development of modern science sanctioned the exploitation of nature, unrestrained commercial expansion, and socio-economic conditions which also perpetuated the subordination of women.

The past few decades have witnessed an enormous concern, development, and growth in both the women's movement and the ecological or the environmental movement. The goals of these two movements are mutually reinforcing, for they both involve the development of world views and practices that are not based on male-biased models of domination. Rosemary Radford Ruether wrote in her book *New Woman, New Earth* in 1975:

> women must see that there can be no liberation for them and no resolution to the ecological crisis within a society whose fundamental model of relationships continues to be one of domination. They must unite the demands of the women's movement with those of these ecological movements to envision a radical reshaping of the basic socio-economic relations and underlying values of this [modern industrial] society.
>
> (Ruether, 1995, 7)

The forces of nature and the strength of the human spirit inspire the novel and poems of Alice Walker. As an activist and an academician, Walker perceived the critical connection between women and nature and believes that writing is the way to correct wrongs that she observes around her. Her recognition of misogyny and exploitation of the environment as parallel forms of male domination undoubtedly takes an ecofeminist stance.

> Walker celebrated the closeness to nature—and specifically to primates—of Black women in particular through revisions of Darwinian and Biblical discourses which though incompatible in obvious respects have each been used in the interests of patriarchy to construct women, animals and people of color as inferior.
>
> (Greenwood, 167)

In an interview with John O'Brien, Walker admitted that she was committed to the cause of Black women but equally concerned about the cause of nature. Her writings are suffused with a concern for the destruction of the pristine glory of the environment. In the Introduction to her work *Anything We Love Can Be Saved: A Writer's Activism*, Walker discloses how her 'activism—cultural, political, spiritual is rooted in her love for nature and her delight in human beings' (1998, 12). She mentions how when she takes a walk amidst nature and sees its beauty, she is reminded of the Black women who 'are flexible like the grass and sheltering like the trees' (111).

II

In the preface to the tenth-anniversary edition of *The Color Purple*, the novel for which she won the Pulitzer Prize, she recognized herself as a 'worshipper of nature.' Walker's work voices the concerns of the marginals who have no voice or significant identity of their own: usually, they consist of though not always, poor, rural Black women:

> Robbed of power and the right to make decisions about their own lives by a range of forces standing against them—their religious leaders; a brutal economic system; racial prejudice, which is often encoded into law; and the frequent misogyny of the men with whom they choose or are forced to share their lives—Walker's women protagonists nevertheless articulate clear visions not just of the wrongs they face, but also of the hope and strength that cannot be quenched within them.
>
> (Donnelly, 2010, 7)

Not every woman character of Walker musters the courage to say, as Celie in *The Color Purple* does to her abusive husband Mr. _____: 'You a low-down dog is what's wrong. ... It's time to leave you and enter into the Creation. And your dead body just the welcome mat I need' (Walker, 1992, 199). However, most of the protagonists develop the ability to stand up for themselves in acts of small or large resistance and confrontation.

The Color Purple (1992) *shows* that the characters of the novel cannot be separated from their physical environment as they endure their separation from nature under constant exploitation led by the dominating males. According to Legler, Walker wanted 'nature to be perceived as more than inert matter ... having a metaphorical status as a speaking, feeling, alive subject' (Legler, 1997, 232). The novel begins with the brutal rapist stepfather's warning words to his elder daughter. 'You better not never tell nobody but God. It'd kill your mammy' (Walker, 1983, 3). Walker uses this shocking opening line to

> narrate the core of the entire story. Under such a painfully cruel silencing brutal and violent act, there is not much left for Celie to articulate herself. Bruised and battered in loveless marriage she starts writing letters to her sister and God venting her innermost feeling to them. In her letters, nature appears as a silent witness as well as a fellow sufferer who accompanies Celie in her journey. In Celie's confessional letters, nature appears to be a visible theme. Humiliated and battered by her experiences, Celie started to feel for everything in nature – the ant, bat, the hoppy toad flattened on the road.
>
> (Greenwood, 2000. 167)

In her narrative Celie expressed her love for flowers, among many properties of nature. Flowers appear as an eye-catching motif in Celie's epistolary narrative which gives her a succour, sustenance, and therapeutic relief. Such motif is metaphorically used by Celie and her sister Nettie. Both the sisters attribute special meanings to flowers in a way to show their renewed spiritual awareness that is reflected in the floral metaphors appearing in their letters. In her earlier novel *Meridian*, Walker documents the struggle of the Black community (especially Black women) and the capacity to resist oppression (both sexual and racial) through self-knowledge. She creates a positive cultural context by portraying the struggle of Black people in an attempt to challenge male hegemony or any other domain of power. Such a struggle holds significance not only for her immediate community but the world at large where individuals/groups are always seeking ways to combat oppression and yet remain whole from within. Thus, *The Color Purple* ceases to be just another feminist text and becomes a universal expression of the oppressed 'other' situated anywhere in the world.

The central philosophy of ecofeminists brings out an integral connection between the body of the female and the core of nature. They find a correlation between mutation of nature and oppression of nature. When nature is annihilated, forests and trees are attacked, it is similar to violence perpetrated on the body of the woman. Shug is a vibrant woman of colour in *The Color Purple* who describes nature as:

> It came to me: that feeling of being a part of everything. Not separate at all. I knew that if I cut a tree, my arm would bleed. I knew just what it was.
>
> (Walker, 1992, 203)

Celie uses the metaphor of flowers, like daisies, while describing her daughter who is forcibly taken away when she was barely two months old: 'I embroider Olivia in the seat of all her daidies. I embroider lot of little stars and flowers too. He took the daidies when he took her. She was ''bout two-month-old. Now she bout six' (Walker, 1992, 15). This visual beauty is sharply juxtaposed against the ugliness, horror, and misery that characterize Celie's life. Her life was marked by physical violence, sexual abuse, and oppression. In *The Color Purple*, flowers are also used as parts of home decoration and clothing. Writing about Harpo's new home with his wife Sofia, Celie says how she makes some curtains out of 'flower sack' (Walker, 1992, 33). Celie in one of her letters to Nettie went on to write about her happiness when she was walking down to Harpo and Sofia's house with 'little red flat-heel slippers, and a flower in [her] hair' (TCP, 196). The flower symbolism in Walker's novel originates from her childhood memories of her mother's love for gardens and her desire to create beauty out of poverty. Towards the end of the novel, nature replenishes itself with bright and vibrant flowers such as lilies, jonquils, roses, and flowering trees. Shug is perceived as a big rose. Through these imageries it becomes evident how flowers represent a peaceful and hopeful life.

The flowers become symbols of joy, freedom, and liberty. Flowers are also integrated into their sense of affordable objects of beauty and sense of fashion and they were meaningful to both Nettie and Celie. Shug's appearance infuses a new life to the trees and stirs Celie out of her apathetic, passive, and unemotional stasis into life and vitality. Shug introduced Celie to the idea of pantheism and eventually she was able to perceive a new meaning of divinity. She was able to discover a spirituality lying in the deepest core of nature and God went on to become a tangible and accessible figure to her. He was closer to her tormented being and she found solace in Him than the conventional patriarchal father figure of divinity. In the course of her journey, Celie like Shug realizes that she cannot find God in the institution of the church or the tenets of the patriarchal religion. She knew that she had to search for him elsewhere. As opposed to an anthropomorphic idea of religion, Walker believed in the philosophy of pantheism, and the novel endorses an idea that God is inherent in all living things within nature. Celie was indoctrinated by Shug's liberal ethos and through her close communion Celie was able to find solace by identifying the presence of God in nature. It was through her assistance that she eventually transcended her liminality and went on to discover her identity and selfhood by accepting the spirituality lying within the heart of nature. By the end of the novel, Celie learnt to fight, resist, and love herself the way she was.

She finally learnt to understand that 'in order to please God, she must learn to love herself since God created her. Loving herself also means leaving Mr._____, accepting the house that she inherited from her mother, and enjoying everything that life has to offer' (Andujo, 2009, 72).

Nature is presented from two perspectives in the course of the novel. It ranges from one extreme – of that of the incapable and powerless in the first half of the novel – to the other, more influential and persuasive force in the second half in the next half. When Celie reasserts her alliance with nature, she is no longer isolated, but eventually becomes assimilated into nature and becomes a part of one ever-growing, fertile relationship.

In Letter 13 written to her sister Nettie, Celie identifies herself with a tree, which is resilient and durable. She attempts to describe the process of insulating herself from the emotional and physical agony that she had experienced as a victim of sexual, physical, and domestic abuse. She also mentioned how she felt that trees too 'fear man,' which subtly hints at Walker's idea of ecofeminism whereby nature and women are both subjected to violence, assault, and destruction. Though Celie felt that she was tough and resilient like the tree, but the 'axe' always posed a threat to her existence. The axe here becomes a symbol of the patriarchal onslaught. The novel introduces us to a form of animism as Walker mentions how some African tribes worshipped the roof leaves for rain and protection. The tree symbolism also refers to this animistic view of the Olinka tribe. Later, in Letter 17, Celie confesses how her psychological transformation, her 'woodenness,' restrained her from showing emotion or feeling affection for Mr. _____'s children. When she pats Harpo and consoles him when he has nightmares about seeing his mother murdered, Celie feels as though she is patting a piece of wood. 'The process of discovering or developing desire begins, for Celie, with the re-appropriation of her own body,' which was taken from her by the world of men – 'first by her brutal stepfather and then passed on to her husband, Albert. The repossession of her body encourages Celie to seek selfhood and later to assert that selfhood through spoken language' (Ross, 1988 70). Later in the course of the novel, after she and Shug are able to find love, companionship, and happiness in each other, Celie's response to nature is completely altered. Her final letter to God is a celebration of trees and flowers in nature as symbols of grace and the marvel of creation. The final redemption happens for Celie when she 'discovers that she is [also a] part of the creation [and] that she fits into the natural order of the world' (Barker, 1999, 61–62).

> Celie with the help of Shug, ... reclaims both her body and spirit, by saving it from domestic abuse, nurturing it, and discovering her sexuality. Sofia Butler, wife of Harpo finds a new home and boyfriend to overthrow the nonchalance, and male chauvinism of Harpo. Squeak overcomes the domination and domestic violence of Harpo and reclaims her voice. Celie and Shug revitalize themselves with Eco womanist awareness and become whole and are ready to face life with renewed s trength. Walker through the novel showcases how nature is the greatest revitalizer of life. She endorses a principle of ecofeminism against the anthropocentric view of man's supremacy over the rest of nature wherein nature is exploited for human benefit.
>
> (Hasanthi, 2015, 161)

The healing and redemptive quality of nature has been highlighted in *The Color Purple*. Nature is not a metaphor in the novel it becomes a living presence. Celie imagines herself as a tree when she defies the claustrophobic domestic violence of Albert. She shares her agony with Harpo, her stepson: 'I say to myself, Celie, you a tree' (TCP, 22). Imagining herself as a tree provides her the strength to combat and confront the violence and oppression. It enables her to offer resistance and accept life in her own terms. In an interview with John O'Brien, Walker mentioned how she did not believe that there exists a God beyond nature. Arguing for a pantheistic perspective of God, Walker has linked issues of environmental pollution, deforestation, and global warming.

Conclusion

One of the primary projects of modern feminism has been to restore women's bodies, which were appropriated long ago by a patriarchal culture, and return it to them. Since the female body is the most exploited target of male aggression, women have learnt to fear or even to hate their bodies. Nature too has been subjected to similar mutilation. According to Adrienne Rich, 'women must overcome these negative attitudes if they are to achieve intellectual progress' (Bloom, 2010, 4). It was through her closeness with Shug who made her realize the spirituality inherent in nature that Celie learnt to accept her body and come to terms with her pain. After her initial revulsion, she finally learnt to accept her femininity which is symbolized as a rose. This 'feminine' representation is further supported by the use of the colour pink which is associated with Shug's dream house. Shug wants to build a house that is 'a big round pink house, look sort of like some kind of fruit' (Walker, 1992, 188). It was through her love for nature that Celie was able to discover her freedom and identity.

For Alice Walker, a womanist who believes about caring for the community, *The Color Purple* becomes the document through which she speaks about the women of the Black community and finally establishes the notion of the necessity of healing in the lives of these oppressed people. Walker looked upon the act of writing the novel as a task which would perform the function of bearing witness so that the community would not forget and learn the importance of love and compassion as pre-requisites for healing. Bearing witness is a phrase which is frequently used in connection with Black literature, especially Black women's writing. It implies the act of faithfully recording all the vicissitudes of the Black racial experience without trying to modify or moderate the traumatic aspects of such experience. *The Color Purple* seems to suggest that Black women by creating empowered sisterhoods can attempt to remedy and rectify the present-day equations between man and woman. Shug's concept of God is significantly a pantheistic one as she tells Celie that one's belief in God is consolidated when she/he feels connected to everything in the world. God, according to Shug, wanted people to note all aspects of the beauty of life and offer praise to Him even if only to admire the beauty of a field of sunflowers. Shug's words encouraged Celie to notice the beauty of the blades of grass, corn fields, and the world around her. Celie imbibed this faith of Shug in the interconnectedness of life which helped her to acquire the faith and courage to stand up to the brutalities of Mr. _____ and finally leave him. Celie learnt to accept her relationship with God when Nettie returned from Africa. In the final letters, Celie goes on to write: 'Dear God, dear stars, dear trees, dear sky, dear peoples. Dear Everything. Dear God' (TCP, 292). An ecological sense of interconnectedness of all aspects of nature and natural life enables her to understand the significance of divinity in nature. At the end of the novel, both Celie and Nettie begin their new journeys of life, respectively, not just as the oppressed other but both of them come of age to resist the oppression and thereby they create new identities and reconstruct the idea of conservative religion to suit their reliance on the philosophy of connectedness of creatures in the world. Through the novel, Alice Walker wants to emphasize how soils, waters, plants, and animals are not commodities that we own but they belong to community to which we belong.

Alice Walker believed that a renewed love, celebration, and joy of the Earth is a worthy ideological pursuit. She emphasized on the change of the role of *Homo sapiens* from conqueror of the land community to plain member and citizen of it. In her letters she repeatedly insists that the earth has become nigger of the world and will assuredly undo us if we don't learn to care for it, revere it, and even worship it. *The Color Purple* brings out the voice of not only Alice Walker the novelist but also Alice Walker the activist who believed, an obligatory need to instigate change. When asked about the artist's responsibility towards change in the perception of nature, she went on to state,

> I can't stand the abuse of the planet and the rampant lack of compassion for the Earth. If you want a world where people are concerned about life on the planet, then you have to be concerned and work for change. But everyone is responsible for the whole creation and the artist has his or her part to do.
>
> (Byrd, Rudolph P, and Alice Walker, 72)

In a poem entitled "Torture," Walker raves against the oppression of the male-dominated society, racial politics, discrimination, assassination, violence, abuse and exploitation, and all other actions of 'scalping the earth' and suggests the only possible answer to counteract all of these.

> When they torture your mother/plant a tree.
>
> ("Torture" *Horses Make a Landscape Look More Beautiful*, Alice Walker)

Works Cited

Andujo, Patricia. 2009. "Rendering the African-American Woman's God through *The Color Purple*" *Alice Walker's The Color Purple*. Ed. Kheven La Grone. Amsterdam: Rodopi. 61–76.

Barker, E. Ellen. 1999. "Creating Generations: The Relationships Between Celie and Shug in Alice Walker's The Color Purple" *Critical Essays on Alice Walker*. Ed. Ikenna Dieke. Westport: Greenwood Press. 55–65.

Byrd, Rudolph P., and Alice Walker. 2010. *The World Has Changed: Conversations with Alice Walker*. New York, New Press.

Bloom, Harold. *Interpretations, Alice Walker's The Color Purple*. New York: Infobase, 2010.

Donnelly, Mary. *Alice Walker: The Color Purple and Other Works*. New York: Marshall Cavendish, 2010.

Gard, Greta. *Ecofeminism: Women, Animals, Nature*. Philadelphia: Temple University Press, 1993.

Garrard, Greg. *Ecocriticism*. London and New York: Routledge, 2004.
Gersdorf, Catrin and Sylvia Mayer. *Nature in literary and cultural studies: Defining the Subject of Ecocriticism: An Introduction*. Amsterdam: Rodopi, 2006.
Glotefelty, Cheryll. *The Ecocriticism Reader*. Athens and London: The University of Georgia Press, 1995.
Greenwood, Amanda. "'The Animals Can Remember': Representations of the Non-Human Other in Alice Walker's 'The Temple of my Familiar.'" *Worldviews*. 4.2 (2000): 164–178. Accessed from **Stable URL:** https://www.jstor.org/stable/43809164 on June 24th 2022.
Leahy, Terry. "Ecofeminism in Theory and Practice: Women's Responses to Environmental Issues." *Journal of Interdisciplinary Gender Studies*. 7.1&2 (2003): 106–125.
Legler, Gretchen. "Ecofeminist Literary Criticism." *Ecofeminism: Women, Culture*. Ed. Karen Warren. Indiana University Press, 1997. 227–238.
Merchant, Carolyn. *The Death of Nature: Women, Ecology, and the Scientific Revolution*. London: Harper, 1990.
Ross, Daniel. 1988 "Celie in the Looking Glass: The Desire for Selfhood in *The Color Purple", Modern Fiction Studies,* Volume 4, No 1, Special Edition pp 69–84. Published by Johns Hopkins University. Stable Url: https://www.jstor.org/stable/26282404
Ruether, Rosemary Radford. *New Woman, New Earth: Sexist Ideologies and Human Liberation*. New York: Beacon, 1995.
Walker, Alice. *From an Interview. In Alice Walker (ed.). In search of our Mothers' Gardens: Womanist Prose*. USA: Harcourt Brace Jovanovich. 1983. 244–272.
Walker, Alice. *The Color Purple*. London: Hachette, 1992.
Walker, Alice. *Anything We Love Can Be Saved: A Writer's Activism*. 1st Ballantine Books ed. New York: Ballantine Books, 1998.
Vakoch, Douglas and Sam Mickey. *Editor's Preface Ecofeminism in Dialogue*. London and New York: Lexington, 2018.
Warren, Karen. "Feminism and Ecology: Making Connections." *Environmental Ethics* 9.1 (1987): 3–20.
Warren, Karen. *Ecological Feminism*. London: Routledge, 1994.

12 Misogyny and Its Postmodern Creation

A Material Ecofeminist Reading of Harold Pinter's Select Women Characters

Saikat Chakraborty

Harold Pinter, the Nobel Prize-winning dramatist, has contributed to the development of political absurdity throughout his career. The term 'Absurd Drama' was coined by Martin Esslin in his book *Theatre of the Absurd*. By the term 'Absurd' Esslin meant the meaninglessness of life in post-War era. The World Wars had shown the world the futility of existence. The idea of questioning the existence or the regular mode of being was a literary movement first brought to limelight by Danish philosopher Søren Kierkegaard. The idea of existentialism and absurdity coexist hand in hand because both the schools emphasize on the chaotic nature of the post-War world. The philosophy of absurdity was concretized and explored with aplomb by dramatists such as Samuel Beckett, Henrik Ibsen, Eugene Ionesco, Eugene O'Neil, and of course Harold Pinter. In this regard, we must emphasize on the fact that Beckett talked about the spiritual absurdity of existence in plays such as *Waiting for Godot* and Ionesco, Ibsen, and O'Neil discussed the absurdity of human relations in plays such as *Amedee*, *The Doll's House*, and *Thirst*. It was Harold Pinter who overtly discussed the political side of absurdity. For Pinter, state domination, agoraphobia, blitzkrieg, claustrophobia, and material exploitation debunk the efficacy of human existence and manifest the ineffectual essence of life. Pinter's political exploration provides us with the canvas to explore the material exploitation of his female characters. Harold Pinter throughout his career had been accused of being a misogynist but we cannot deny the fact that with the aid of his inversion technique he critiques the material exploitation of women by being a misogynist himself. Kimball King believes that 'Pinter has brought a form of natural speech to the stage that has surpassed most of the ambitious attempts of his predecessors' (King 244) His plays juxtapose 'approbation,' 'blasphemy,' 'gluttony,' and 'buggery.' The characters lack unity and they are portrayed as voids, perhaps to represent the emptiness of the contemporary culture.

The prowess in language has always been Pinter's subject as he claims, 'I am pretty well obsessed with the words when they get going.' Pinter defines language or speech as the 'stratagem that covers nakedness' (Esslin 46).

DOI: 10.4324/9781003383970-15

Pinter's characters in the plays do not share a pleasant relationship because they always use language not only to communicate but to dominate each other. The idea of using language as a perpetual power system can be explicated with Foucault's commentary on Structuralism. Foucault in his book *The Order of Things* argues, 'Structuralism to be the awakened and troubled consciousness of modern thought' (Foucault 226). For him, language and its perpetual power structure regulates with certain social discourses and social codes. On the other hand, Pinter's characters with their usage of language seem to engage in Wittgenstein's proposed argument about the language game (Wittgenstein 3). According to Wittgenstein,

> The language is meant to serve for communication between a builder A and an assistant B. A is building with building stones: there are blocks, pillars, slabs and beams. B has to pass the stones, and that in the order in which A needs them. For this purpose they use a language consisting of the words 'block', 'pillar', 'slab', 'beam'. A calls them out; B brings the stone which he has learnt to bring at such-and-such a call. Conceive this as a complete primitive language.
>
> (Wittgenstein, 3)

However, in Pinter's plays this primitive idea of language does not work in its conventional form. The use of language becomes multifarious and we see the whole idea of communication being toppled by a sense of domination and manipulation. In this regard Wittgenstein suggests,"

> A game consists in moving objects about on a surface according to certain rules ..."—and we replied: You seem to be thinking of board games, but there are others. You can make your definition correct by expressly restricting it to those game.
>
> (Wittgenstein, 3)

These political overtones, language games, and of course the machinery developments contribute to the suppression of women as well as nature in the postmodern era. The post-modern era is replete with scientific developments, simulation, and hyperreality. Scientific developments of course naturally lead to exploitation of nature. Sociologically speaking, Pinter presents in front of us a world of globalization. The bizarre atmosphere of the plays brings into context the deteriorating human values and the atrocities of globalization. Anthony Giddens talks about how our progression from agrarian society to industrialized society is actually retrogression. According to him, our progression from agriculture to industry has actually given us atrocities, jealousy, competition, and vehement attitude. So, this is a 'retrogression of human civilization' (Giddens, 34). This sense of retrogression leads to the exploitation of nature as well as women. Therefore, the globalized post-modern world generates a sense of Deleuzean 'desire' (Holland,

53). For Deleuze, desire 'is the production of reality,' as Holland points out, 'desire produces and its product is real' (Holland, 54). Therefore, we understand that the idea of desire is related to the idea of 'production' in Marxian terms.

While talking about this Deleuzean notion of desire (Holland, 53) we must understand the fact that post-modernist views have a strong post-Marxist precept and its presence is ever pervasive (Laclau and Mouffe, 9). Post-Marxism for Laclau and Mouffe is a transition from Marxism and the change that occurs through this transition is not only ontic but also ontological (Laclau and Mouffe, 10). Put simply, the economic value of a globalized world – that cannot be determined by the Hegelian and naturalistic paradigms of Marxism – can actually be interpreted through this ontological shift. So, the characters in Pinter's plays are in relation to each other in terms of economic value. They all live in a world that Baudrillard calls 'brothel of capital' (Baudrillard, 9). As Baudrillard suggests,

> This is the generalised brothel of capital, a brothel not for prostitution, but for substitution and commutation. This process, which has for a long time been at work in culture, art, politics, and even in sexuality (in the so-called 'superstructural' domains), today affects the economy itself.
>
> (Baudrillard, 9)

Therefore, sexuality is also determined according to the economy and material exploitations. The 'brothel of capital' creates a panacea for materialistic exploitation, especially of women.

Now to comment on the material exploitations of nature, women, and their interrelationships, we have to discuss the genealogies of the movements like ecology and feminism. According to Carolyn Merchant, the ancient identity of nature as a maturing mother links women's history with the history of the environment and ecological change. The female earth was central to the organic cosmology that was undermined by the Scientific Revolution and the rise of a market-oriented culture in early modern Europe. The ecology movement has reawakened interest in the values and concepts associated historically with the pre-modern organic world. The ecological model and its associated ethics make possible a fresh and critical interpretation of the rise of modern science in the crucial period when our cosmos ceased to be viewed as an organism and became instead a machine. Both the women's movement and the ecology movement are sharply critical of the costs of competition, aggression, and domination arising from the market economy's 'modus operandi' in nature and society. Ecology has been a subversive science in its criticism of the consequences of uncontrolled growth associated with capitalism, technology, and progress –concepts that over the last two hundred years have been treated with reverence in Western culture (Merchant, Introd. xx). Thus, it is inevitable that ecological disaster due to the blitzkrieg phenomena in the post-War era and the exploitation of women due to the scientific advancements and the materiality of the market are indispensable regarding their association to each other.

Now, while talking about Pinter's characters, at first, I would like to discuss the character of Meg in one of his most renowned play *The Birthday Party*. The character of Meg in the play signifies the mother figure of the play. She is the one who signifies the anxiety of the natural world and of women as well. She is the embodiment of the cosmos so to speak. However, the domination of 'Cassandra School' over 'Cornucopian School' is well embodied with the manifestation of the oedipal desire embedded within Stanley Webber. The Cornucopian school emphasizes on the Mother Nature's omnipresent prowess to outweigh the human exploitation (Jonsson, 153). However, Cassandra school or what Jonsson calls 'Malthusian misery' talks about the destruction of ecology where even Mother Nature is helpless and is robbed of her agency by human beings (Jonsson, 165). The term 'Malthusian Misery' is derived from *An Essay on the Principle of Population* by Reverend Thomas Robert Malthus. The idea of nature as motherly figure is very important in this context because the exploitation of the mother in the hands of her own children is well manifested by the oedipal complex shown between Stanley and Meg. Stanley has never been depicted as the son of the family. Instead, he has been shown as an expatriate. Pinter does this intentionally not only to emphasize on the anxious wanderings of the Jews to save themselves from the Nazis but also to contribute to the post-modern manifestation of unreliable narrator made famous by Fowles in his seminal work *French Lieutenant's Woman* and the illusory solidarity of the past. However, as the relationship is shown in the play, Stanley can easily be concluded as the philosophical extension of the son of the family, if not the son himself.

The oedipal complex is visible in the play from the very beginning. In a conversation between Meg and Stanley, we find:

> *Meg:* Stan?
> *Stanley:* What?
> *Meg (Shyly):* Am I really succulent?
> *Stanley:* Oh you are. I'd rather have you than a cold in the nose any day.
> (Pinter, 1991 Act1 19)

The idea of the male desire to exploit and consume the female body is very conspicuous in these lines. The women are subjected to exploitation in the post-War era not only in the hands of the state and mechanical developments but also in the hands of their so-called near and dear ones.

As the play progresses, the idea of organized violence to exploit nature in the hands of the state is depicted in the beginning of the Act2. Here we find McCann – the representative of the state machinery – seen to be tearing a sheet of newspaper into five equal strips. The five equal strips signify the effect of organized violence towards nature. Here the newspaper becomes the representative of the natural ecology and the tearing of it signifies the exploitation of nature in the hands of the nation state. The play moves forward hastily and the idea of this oedipal desire for the female body and state's dominance over the natural resources are almost linked together where we see the characters playing 'Blind man's Buff' to celebrate Stanley's birthday. The game is a signifier of the unruly exploitation of nature and female body in the hands of the state and its machinery prowess. Therefore, even the mother's womb is invaded in the age of materiality and mechanical reproductions.

In the game, Stanley is blind-folded and it has two-fold interpretations. Firstly, the robbery of his eyesight that presents him as the Oedipus himself and Meg becomes the representative of Yocasta. Secondly, Stanley is the signifier of the offspring of the woman womb (Meg) and the robbery of his agency by Goldberg and McCann, the representatives of the state, shows how machinery of a nation state even exploits the mother's womb. The drum used to create the cacophony also signifies plethora of interpretations. The drum with its round shape stands as a representative image of the mother's womb. Stanley's entanglement with the drum shows his desire to go back to the mother's womb to escape from stately exploitations. This signifies the human desire to escape existential crisis and go back to the mother's womb. This is what Bowlby refers to as 'return-to-womb craving' (Bowlby, 357). He aptly quotes Melanie Klein and suggests, 'This mental and physical closeness to the gratifying breast' in some measure restores, if things go well, the lost prenatal unity with the mother and the feeling of security which goes with it (Bowlby, 357). However, Bowlby suggests, the secondary drive of sucking and clinging is completely different from the desire of a return to the womb and it is according to him 'both redundant and biologically improbable' (Bowlby, 351). Therefore, Stanley's helpless entanglement with the drum is a signifier of perpetual crisis and biological improbability of the fantastical return.

Thus, we see Stanley, while moving towards Meg, gets entangled with the drum and staggers along his walk towards Meg. The scene presents in front of us the exploitation of female body as well as nature by state and the market economy that even invades the mother's womb. The staggering image represents the struggle of nature and women during their confrontation with post-modern machinery. The drum and its precarious condition represent the helplessness of the mother's womb.

Apart from this, Pinter entangles existential crisis with the sense of exploitation to exemplify this idea of exploitation that even invades into the life of a mother's children vis-à-vis materialist and market-oriented economic exploitation of human beings. Here the idea of the cornucopia is re-asserted as nature is considered as the mother, metaphorically represented by Meg and the natural resources metaphorically represented by Stanley. There is no denial of the fact that most of Pinter's plays are a resurfacing of his anxiety as a Jew due to the threat of detention camp and execution created by the Nazis. So, the same anxiety is represented in the character of Stanley. In a conversation, we find Stanley saying:

> *Stanley (advancing):* They're coming today. They're coming in a van.
> *Meg:* who?
> *Stanley:* They've got a wheelbarrow in that van.
> ...
> *Stanley:* They're looking for someone. A certain person.
>
> (Pinter, 1991 Act1 24)

The lines strongly exemplify Stanley's anxiety that eventually leads to his take-away by Goldberg and McCann. As the play progresses, we find Stanley's glasses have been broken:

> *Goldberg:* What's the matter with your glasses?
> *Goldberg bends to look.*
> *They're broken. A pity.*
>
> (Pinter, 1991 Act3 81)

With the unfolding of events, we find that it is the stately prowess of Goldberg and McCann that eventually robs the child from Mother Nature (Meg) and the child is taken away for the market exploitation. Here, the character Petey becomes the embodiment of criticism against such stately exploitation of mother and her child. He's heard saying:

> *Petey (broken):* Stan, don't let them tell you what to do.
>
> (Pinter, 1991 Act3 86)

Here the character Petey works as the manifestation of post-War Prometheus who becomes the harbinger of life and posits himself in sharp contrast with the machinery exploitation of stately prowess. Here Petey also critiques Kafkaesque interrogation and shows the importance of taking a stand against such exploitations.

The other female character in the play, Lulu, is an exclusive representative of how a female body is exploited by market economy, state colonization, and post-modern machinery. Carolyn Merchant argues that women due to their association with the idea of fertility seem to represent nature itself (Merchant, Introd. xxi). So the character of Lulu is not only a representative of the women but also the nature in microcosm. We find her saying:

> *Lulu:* He was my first love, Eddie was. And whatever happened, it was pure. With him! He didn't come into my room at night with a briefcase!

Goldberg: Who opened the briefcase, me or you? Lulu, schmulu, let bygones be bygones, do me a turn. Kiss and make up.

...

Lulu (with growing anger): You used me for a night. A passing fancy.
Goldberg: who used you?
Lulu: You made use of me by cunning when my defences were down.
Goldberg: who took them down?
Lulu: That's what you did. You quenched your ugly thirst. You taught me things a girl shouldn't know before she's been married at least three times.

(Pinter, 1991 Act3 80)

In these lines it is very clear that Goldberg, the representative of the state machinery, has exploited Lulu sexually. The exploitation not only is a representation of women's commodification in the market economy but also shows the overreacting effects of the World Wars on nature. The use of modern technology, phenomena such as blitzkrieg, had destroyed the world ecology completely.

The Homecoming is a play that deals with the postmodernist element of 'fall of the metanarrative' proposed by Jean-François Lyotard (Powell, 29). Here family is a metanarrative. The play revolves around a family that consists of only a single woman character, Ruth. She is the wife of Teddy, the philosopher and the elder son of the family. The play represents a homecoming of Teddy but eventually it turns out to be a homecoming for Ruth because the play portrays a time frame when post-Marxist (Laclau and Mouffe, 9) economy had engulfed women agency as well as nature. Ruth's homecoming can also be viewed as a desire of Ruth to become the mother of the house and also the wife of Teddy. But this Deleuzean (Holland, 53) notion of desire seems to change its territorial axis. At the end of the play, rather than becoming a wife, Ruth turns out to be a prostitute.

The play pervasively talks about opposing establishments or metaphysics of presence (Derrida, 88). Ruth, for example, disrupts the traditional structure of a family. She leaves her own family and joins the family of his so-called husband only to serve them as a prostitute. With this very idea, Pinter actually disrupts the idea of a family, 'a metanarrative' (Powell, 29), that has fallen and therefore Pinter justifies the idea of 'fall of the metanarratives' of Lyotard. It not only facilitates fall of the metanarrative but also speaks a lot about little micronarratives where every character seems to be justifying the different functions (Powell, 32), thereby positing a true post-modernist multiplicity. This breakdown of the family is consolidated and justified in the second act when Teddy comes downstairs to leave but Ruth accepts Lenny's offer to dance with him and she allows him to kiss her. The disturbing elements and the disrupted condition of the family gets densified when Joey sits on the sofa with Ruth and makes out with her in front of her husband Teddy who behaves shockingly casual. Baudrillard in his book *On Seduction* talks about this idea of seduction and according to him seduction is feminine. Sex, on the other hand according to Baudrillard, is masculine, centred on the phallus, natural, and non-artificial (Powell, 65–68). Freudian hypothesis, according to Baudrillard, is perfect, where it advocates that libido is actually determined by masculinity. The consumption of Ruth's body by the male characters seems to justify Baudrillard's notion.

The play seems to invoke the Deleuzean notion of 'assemblage' (Wise, 77) as the characters are in relation to each other in respect to their territory and are in a constant phase of folding–unfolding and becoming. While exploring the characters we understand that Ruth's territorial axis completely changes when she arrives in England from America and rather than being a mother and a wife, she becomes a prostitute. This territorial axis of an assemblage is regulated in the play with the aid of economic determinism. Teddy's futility as a source of income becomes the reason of his departure from the so-called 'home' and Ruth's utility as a prostitute results in an actual homecoming for her, as if the role of a prostitute was chasing her all the time. Sam and his territorial axis are only present in the society because he is employed and he also has this desire to become the head of the family. On the other hand, due to his economic stability Max becomes a pathetic character who struggles to stay in his own family and here, rather than being in assemblage to the other characters, Max's desire to become the head of the family comes to the foreground. This notion of desire (Holland, 55) extends itself to the extremes, at the last scene of the play. Here Deleuze suggests,

> the function of social representation is precisely to separate desire from reality (to separate a body from what it can do, as Spinoza characterizes the effect of superstition), to retrospectively inject so-called 'needs' and scarcity and lack into a desire reality relation that is immediate and full on the unconscious, species-being level. The result is that individuals and groups come to believe consciously that they lack or need something: a something that had in fact been produced by desire itself, but that subsequently gets taken from them by social order.
>
> (Holland, 54)

We find Max suffering from the lack of a woman in the household and therefore generates a desire for that, but Ruth's transformation from the mother to the whore manifests the social taking away of the object of desire within the economic value system. Thus, we see Max kneels down while saying to Ruth:

> *Max:* I am not an old man.
> *Pause*
> *Do you hear me?*

He raises his face to her
Kiss me.

(Pinter, 1965 Act2 82)

So, it is evident that economic determinism is the only machinery that contributes to the exploitation of Ruth as a site of commodification. This idea of commodification and economic determinism eventually leads to the fall of social sanity and conventions. Thus, I would like to conclude by saying that Pinter's female characters and their representations provide us with a greater vista for an ecofeminist and materialist study in order to critique the post-War era as well as the contemporary times.

Works Cited

Baudrillard, Jean. *Symbolic Exchange and Death*. Sage Publications, 1993.

Bowlby, John. "The Nature of the Child's tie to his Mother". *International Journal of Psycho-Analysis*, Vol. 39, 1958, pp. 350–373.

Derrida, Jacques. "Structure, sign, and play in the discourse of the Human Sciences". *Modern Criticism and Theory*, edited by David Lodge. Dorling Kindersley Pvt. Ltd, 2007, pp. 89–102.

Esslin, Martin. *Pinter: A Study of His Plays*. W.W. Norton & Company, 1976.

Foucault, Michel. *The order of Things: An Archaeology of Human Sciences*. Routledge, 1966.

Giddens, Anthony. *Sociology*. Polity Press, 2006. (www.scribd.com.mob>doc).

Holland, Eugene W. "Desire". *Gilles Deleuze Key Concepts*, edited by Charles J. Stivale. Acumen Publishing Limited, 2005, pp. 53–64.

Jonsson, Fredrick Albritton. "The Origins of Cornucopianism: A Preliminary Genealogy". *Critical Historical Studies*, Vol. 1, No. 1 (Spring 2014), pp. 151–168.

King, Kimball. "Harold Pinter's achievement and modern drama". *Pinter at 70*, edited by Luis Gordon. Routledge, 2001.

Laclau, Ernesto and Chantal Mouffe. *Hegemony and Socialist Strategy: Towards a Radical Democratic Politics*. Verso, 1985.

Merchant, Carolyn. *The Death of Nature: Women, Ecology and The Scientific Revolution*. Harper and Row, Publishers, 1985.

Pinter, Harold. *The Birthday Party*. Faber and Faber Limited, 1991.

Pinter, Harold. *The Homecoming*. Grove Press, 1965. (https://kupdf.com).

Powell, Jim. *Post Modernism for Beginners*. Orient Longman, 2001.

Wise, J. Macgregor. "Assemblage". *Gilles Deleuze Key Concepts*, edited by Charles J. Stivale. Acumen Publishing Limited, 2005, pp. 77–87.

Wittgenstein, Ludwig. *Philosophical Investigations*, translated by, G.E.M Anscombe. Basil Blackwell, 1953.

13 The Unnatural Nature
Edgar Allan Poe and Eco-horror

Riman Rakshit

The question of what lies outside the boundaries of our known domestic sphere has haunted humanity for centuries. Every culture has a recorded fear of what creeps in the dark, the wilderness, and in what ways might 'it' invade and colonize us. Our folklores tell graphic tales of *Jokkho* and *Nishi* hiding in the darkness, waiting to prey upon us. Contemporary culture has seen an increasing fascination with stories about the violence inflicted by Nature. Social media dramatizes natural disasters as both a warning and a beckoning for humans to change our unruly ways of dominating Nature. Perhaps no other time has been more suitable for looking into the metaphysics of eco-horror and eco-phobia than the pandemic which has led to an evident loss of our superiority and our autonomy. The narratives of the pandemic have resulted in a revision of eco-horror, a fear that lurks both within the domestic sphere and without. With faces being replaced by masks and shields, there has been an evident rise in anxiety to define and express ourselves. Masks not only make us seem identical to one another, but they also limit our expressions; we feel stifled and caged in them. That an unseen 'other' is spreading within our known communities like wildfire killing millions of people has made humans doubtful of our position in the natural hierarchy. The virus is invading our domestic spaces, our personal spaces (bodies), and manipulating them to betray us. The fear of what lies in the future is increasing our anxiety and animosity towards nature. Forest fires raging in the Amazon, bright red skies in California and Indonesia, are a few of the other incidents of the recent eventful years which have made us wonder if it is indeed the beginning of a natural apocalypse that Hollywood potboilers would want us to believe. Bernice M. Murphy argues that, – by confronting readers and viewers with the realization that – 'absolutely catastrophic ecological disaster looms not in the distant future, but is in fact already unfolding,' the idea of eco-horror has been 'assimilated into popular culture' (179). However, the validity of such horror is not just relevant to the macrocosm of humanity, but equally important to the microcosm of individual struggles in making sense of the dangers of Nature. For we can only understand what we know, and facing the fear lurking beneath the unknown also implies coming to terms with what we do not know, and this

DOI: 10.4324/9781003383970-16

is the primary seat of our fear. The idea of our surroundings slipping away from our control and turning hostile against us is a horror that perpetuates out of our collective guilt of colonizing Nature, which is not too different from that of the English *Babus* who would have vivid fearful dreams of their Indian servants lurking outside in the *verandah*, and invading their bedrooms and murdering them in their sleep, in a colonial India.

The Tradition of Eco-horror

In his seminal work, "Theorizing in a Space of Ambivalent Openness: Ecocriticism and Ecophobia," Simon Estok claims that there is 'a need to talk about how contempt for the natural world is a definable and recognizable discourse' (204). He names this 'ecophobia' and argues that for ecocritics, 'ecophobia is an irrational and groundless hatred of the natural world, as present and subtle in our daily lives and literature as homophobia and racism and sexism' (208). He calls upon ecocritics to recognize the nature of repulsion and alarm that characterizes people today. Estok's observations begin a line of conversations on the nature of such fear, analysing the various metaphors for eco-phobia and the deep roots of the horror entrenched in our collective subconscious over the uncanny in Nature. The 'Uncanny' (*Unheimlich*), Freud writes, is 'that class of the terrifying which leads back to something long known to us, once very familiar' (1–2). The uncanny is the seat of the repressed which has emerged into the light. Thus, whereas the word *heimlich* has the dual senses of homeliness as well as concealment; *unheimlich* is the forgotten, the hidden, emerging out of the shadows, into the known. The tradition of eco-horror has its roots in British settlers finding themselves in distant lands constituted by different sets of flora and fauna. This difference created settlers' anxiety by which dissimilarity became a signifier of the ghastly (uncanny). Bernice Murphy has commented on the acute eco-phobia experienced by 'white settlers who set out to create a "new world" in the midst of a vast, unfamiliar, and often physically treacherous landscape already occupied by resentful native inhabitants' (181). Against this tradition arose another, centred on 'white ego' – the eco-philia of the transcendentalists who argued that Nature is a mirror which reflects the human soul. Adapting Emerson's theory of Nature, they emphasized on self-love as a cure for eco-phobia, to transform the foreign terrains into manicured lawns would cure the fear, since it would eliminate the dissimilar.

What Emerson famously called 'self-trust' laid the foundation for his human-centred concept of nature; he envisioned Nature as a reflection of the human and demanded that we learn 'to worship the soul' and as Sara L. Crosby comments in her essay "Beyond Ecophilia: Edgar Allan Poe and the American Tradition of Ecohorror"

> see, that nature is the opposite of the soul, answering to it part for part ... This love of nature as love of self completes both (John) Winthrop's

> and (John) Smith's program to assimilate a frightening environment through the exaggeration of the ego.
>
> (Crosby 2014, 4)

For centuries, through oral and written narratives, the fear has been passed from one generation to another in coded semantics which signifies the unnameable in Nature. Gothic fiction as a genre pushed the boundaries of the unknown, the impossible, the macabre, and it is not a coincidence that many iconic Gothic tales such as *The Castle of Otranto* or *The Monk* are situated in the distant past, locked in a temporal bubble. However, ironically, the reason why these tales are so effective is that fear has always been a 'present' concern. Perhaps no other could make this terror as palpable as the American writer Edgar Allan Poe. The urgency in his writings, the recurrence of metaphors for madness, illness, incest, mirroring, and above all a 'sinister' Nature looming on his subjects, makes Poe so quotable and relevant today when more ecocritics are calling for action and practical solutions in the academic area. For Poe, the natural world was far bigger, inhuman, and indifferent. In his nature sketch, *The Island of the Fay*, he regards 'dark valleys,' 'grey rocks,' 'mountains,' etc. 'as themselves but the colossal members of one vast animate and sentient whole … whose cognizance of ourselves is akin with our own cognizance of the animalculae which infest the brain' (600 quoted in Crosby, 6). He disagrees with the traditions of eco-horror and eco-philia and does not sympathize with his male protagonists who control and oppress both their environments and the women of their families. Poe's narratives, on the other hand, are a celebration and a prophesy for the return of the repressed, as the wilderness rebels and destroys the oppressive structure that controlled it. In stories such as *Berenice* and *The Fall of the House of Usher*, the victory of Nature can be equated to the victory of the woman (having their bodies controlled and defiled) who re-emerges in a ghastly state, yet in her deformity she claims her autonomy, and lets herself be heard (Berenice) or seen (Madeline).

Poe created the binaries of Man/Woman, Man/Nature, Love/Fear, Knowledge/Unknown, Self/Other, Sanity/Madness, Rational/Irrational, in his macabre tales, and as he subverts these binaries, the (male) characters, as well as the audience, are unsettled. For just like the readers, the characters are also trying to rationalize a world which they gradually realize to be beyond their understanding and control. The lack of Knowledge and Power emasculates them and spirals them into madness and despair. The ones who survive are the ones who have to flee the collapsing order of Nature and Reality around them. Thus, the Fear itself takes a visceral shape of its own, altering the identity and the lived reality of the subject experiencing it. The battle against the Other for power then becomes a schizophrenic battle between two halves of the Self – the rational and the irrational fighting against each other.

The Space of Fear

The idea that space is a passive component, serving as an empty vessel for social interactions, has been negated by human geographers who emphasize on the importance of space as an active participant constructing the identities of the people inhabiting it, as well itself being re-constructed and re-organized. Human geography is the 'study of people and their activities from spatial and ecological perspective' (Husain, 520). Henri Lefebvre in his book co-authored with Donald Nicholson-Smith, *The Production of Space* elucidates that a space is socially produced through the acts of its inhabitants:

> human beings do not stand before, or amidst, social space; they do not relate to the space of the society as they might to a picture, a show, or a mirror. They know that they have a space and that they are in this space ... they act and situate themselves in space as active participants.
> (Lefebvre 2009, 294)

This social production is not an innocent act, as different power relations enable a spatial hierarchy. This is evident in the domination of the city on the surrounding countryside. The city, symbolic of rationality and power, becomes the nerve centre of the self, which gradually expands and infiltrates the country as the latter loses its identity. The countryside becomes symbolic of the repressed which threatens to rebel against the urban obsession with order. In many of Poe's Gothic tales, we encounter 'the anti-modern rural flaneur ... who, as a product of modernity, of urban modernity, takes the position and challenge of the flaneur and wanders in and through "landscape"; responding to "landscape" as if it is the modern urban imaginary' (Grimshaw, 3). The enigmatic narrator in *The Fall of the House of Usher* is such a flâneur whose detached yet ominous commentary of the deteriorating ecology around the House induces the threat in untamed nature:

> I looked at this scene, I say, with a complete sadness of soul which was no healthy, earthly feeling. There was a coldness, a sickening of the heart, in which I could discover nothing to lighten the weight I felt. What was it, I asked myself, what was it that was so fearful, so frightening in my view of the House of Usher?
> (1)

Even before the entrance of the titular characters and the anatomy of the house is revealed, the narrator's depiction of the rural space as wild, disorderly, stagnant which contrasts to the conception of urban space as controlled, orderly, and vibrant. This dichotomy arouses fear within the narrator as well as the reader, both of whom have been indoctrinated with an urban viewpoint.

Furthermore, Space constitutes the mental space – the intimate thoughts and ideas hidden from the outside world, the physical space of the body, the domestic space, and the space of the vast unknown. The mental space is the most intimate seat for ideas and emotions, and the fundamental unit of private space, with a significant area inaccessible and repressed in conscious performances. Naturally, when this space feels invaded by a foreign agent, it causes paranoia, as is the case with the narrator of *The Raven*. The prophecy of the bird's single-worded answer threatens the stability of the narrator's psyche, for he becomes aware that the uncanny (his repressed thoughts and actions about his beloved) will never disappear; it will loom on his subconscious, breaking the borders he had erected in his mind.

The secret of the domestic (incestuous) space in *The Fall of the House of Usher* contaminates the landscape around the house which also projects the sick mental and physical spaces of its inhabitants. Like the narrator of *The Raven*, Roderick Usher too is fearful of his curated, controlled domestic space being infiltrated by the unknown. Madeline Usher's physical space (body) being violated and dominated by her brother to an extreme end (being buried alive) is a signifier of male tyranny over Nature. The (uncanny) Madeline can only reclaim her space after she destroys her brother and becomes one with Nature.

In this chapter, I shall analyze how eco-horror functions in the works of Edgar Allan Poe to construct and manipulate the subjectivities of the characters as well as to subvert the expectations of the readers. I have considered two iconic tales to reflect on the dual acts of subversion and manipulation – *The Raven* and *The Fall of the House of Usher*.

The Nature of Fear

In the middle of a December night, hearing a gentle rapping on his chamber door, the speaker of *The Raven* begins to vividly imagine what might lie in the darkness outside his chamber. The greater the depth of his imagination, the more is he filled with *fantastic terrors*, echoing how the imagination of the Other itself becomes a concrete object of fear. The darkness coupled with the cold has heightened his anxiety of the Unknown in Nature waiting to invade his domestic space as he quietly prays for morning to come. He rationalizes that it might just be a visitor, as we often tend to seek meaning for events we do not apprehend. However, on opening the door his fear is elevated on seeing the darkness, an apt weapon for the Other to maintain its anonymity. As he stares into the dark, the silence seems unbearable, he wonders about the evil creeping in Nature, and he dreams about things no human can imagine. This is the initial step of wild Nature terrorizing the educated human (for he was indulging in 'many a quaint and curious volume of forgotten lore'), using his imagination as a weapon against him. As Estok observes:

> the unpredictability of Nature is one of the biggest roots from which Ecophobia grows. In the Western Civilisation, the emphasis on

> controlling and domesticating Nature has been so highlighted, that humans think it is within their inherent rights to have Power and Control over the Environment.
>
> (210)

Yet Poe living in an early nineteenth-century America, suffering from the afterbirths of a Nation, knew through personal experiences and memories that Nature cannot be controlled. His writings almost seem a mock response to the hubris of Westerners, showing them that in the end humans are powerless and only a cog in the giant wheel of Nature. In that, Poe can be seen, not as a writer who portrays classic eco-horror of the whims of 'bad Nature' (for as Estok comments, Nature is 'morally neutral'), but as one who is mocking Western Egotism (Crosby, 518).

The speaker's loneliness and a sense of needing closure over his lost love 'Lenore' intensifies his fear, thus reflecting that even inside the comforts of his known chamber he is not secure. He whispers the name of his lost love and the darkness echoes it back, learning more about him. This is one of the fundamental seats of discomfort, the realization that Nature will always have an upper hand since she knows more about us than we can ever know about her. Our known surroundings can turn hostile against us and even our body can seem foreign in her presence. As the sound of a *tapping* makes him equally curious and fearful, he ventures to check the source deciding it to be only the wind. This assumption is briefly broken as he opens the window and a raven flies in. The bird is a harbinger of agony and despair and as the wild ones in Alfred Hitchcock's *Birds*, the raven is a symbol of insidious nature which is a catalyst in the tragic downfall of the protagonist. In his paper, "Ecocriticism in an Age of Terror," Estok notes that tragedy measures out both human impotence before nature and a persistent inability to conquer, subdue, and maintain control over nature (4). The speaker knows that the Raven is capable of uttering a single word 'Nevermore,' and yet he cannot stop asking questions out of morbid curiosity and anthropocentrism. The more personal the question, the darker the meaning behind the one-word answer, and the speaker's despair increases until he descends into ravings, calling the Raven a *fiend*. Even after realizing that the Raven is capable of only uttering that word, the speaker's innate narcissism forces him to make the word about himself and his agonies. This confirms Estok's hypothesis that Nature does not take notice of perceived privileges human beings regard as their rights:

> The whole question of human exceptionalism emerging through Sandy or Sendai or any number of other natural disasters tells the world that we humans — the whole bunch of us — are nothing.
>
> (3).

It is humans who imagine that they have an agency to overpower and subjugate the environment. The realization of the speaker that the Raven will never answer to his satisfaction sends him into frenzy where he calls the bird – a force of evil. Yet, the only crime the bird can be accused of is its limited vocabulary.

It must be noted that Poe chose the Raven, not because of its morbid nature, but because it is parrot-like (Farrant and Urakova, 161). Poe surely subverts the readers' expectations when he claims that he chose the Raven while creating the poem, not because it is a harbinger of ill omen but because it is a talking pet. By overturning the master/pet relations, Poe illustrates who the real master is. However, for the readers who are unused to seeing stately Ravens at their homes, Poe not only reverses a pet/master dichotomy, with all its recognizable racial implications, he also makes the raven absolutely significant by abstracting it from the natural habitat (162). Thus, the ordinary becomes uncanny. Instead of being in Nature, the Raven breaks conventions to fly indoors. The uncanny in Nature is propagated through an overturning of conventions. It is important to note that the bird is perched on the bust of Pallas, a symbol of wisdom, and yet it is uttering the same jargon albeit prophetically. Perhaps suggestive of how useless is man's knowledge when confronted with the *unheimlich*. Poe's homeliness is one aspect of Freud's uncanny – of the *heimlich* becoming unfamiliar – the mindless bird delivering absolute truth, by dint of retaining its unconscious animality (165). To both the subject and the audience, the overturning of traditional space for wild Nature might be ominous, but to Poe, both are expected of the bird.

The Raven represents his guilt which he had repressed in his subconscious. And through each of its answers, a new horrific realization dawns on the speaker. It seems as though he has lost his subjectivity to the bird, until he is no more than the shadow on his chamber floor devoid even of his bodily functions. His suppressed guilt and loneliness over the tragic death of his love unleashes and breaks his sanity for which he blames Nature, although it was, he who uttered the questions after being manipulated by the presence of the stately raven who seemed to him like a prophet. Too late does he realize that the Raven may not be a prophet at all. The horror here lies then not on a malicious Nature invading the domestic quarters of the human, but Nature holding a mirror to his psyche which will bring out answers he wished he had not known, an ironic application of the transcendentalists' idea of nature showing the best of oneself. The Raven being eternally in the room is a reminder to the speaker of his loss of identity and his morphed subjecthood, where he can no longer hide from his past or from Nature, just like Coleridge's albatross hanging around the mariner's neck is a symbol of his guilt and misdeeds from which he cannot seek refuge.

Indeed, Poe subverts the Romantic image of the transcendent, spiritual, almost bodiless bird, the 'blithe spirit,' and he does it by both 'domesticating' and 'estranging' his raven, making it more like a parrot or a pet, yet an uncanny pet, a device of self-torture and the very opposite of sweetness and obedience (Farrant and Urakova, 163). It is interesting to note that Poe deliberately did not cast his characters suffering from eco-phobia in a sympathetic light. He meant them to be seen as the self-centric narcissists that they are, in their comfort on materiality and bookish knowledge. This is most visible in the Gothic tale *The Fall of the House of Usher*, where the landscape, the house, and the characters have morphed into one bleak reality. According to Estok:

> If we understand ecophobia as the imagining and marketing of badness in Nature, then we can see also that it is a central element of anthropocentrism, perhaps not the sole trait characterizing human relationships with the natural world, but a very important one nevertheless. Ecophobia textures humanity's relationship with the natural world.
>
> (4)

Roderick Usher's inability to be the 'superior' in the binary of Nature/Human defines his tragedy. In his vanity, he refuses to tend to Nature, to find out the reason why the natural landscape is degrading, the tarn getting more polluted and darker every day, and instead focuses his time and energy to better his health and mind not realizing that he is only a part of his environment and cannot be cured independently. His phobia is rooted in his self-centredness.

Two different strands of thoughts have emerged on this issue – the first is the classic eco-horror take by Rodney Giblett who has suggested that the toxic wetlands poisoned Usher and the mansion with its chiasmic evil thus leading Poe to illustrate the evils of wild Nature, and the second belongs to Sara L. Crosby who has seen it as a cautionary tale for 'rich aristocrats' like Roderick Usher, to not destroy the natural environment by considering it their private domains, and poison Nature with their sullen tempers and their ugly, depressing mansions (Crosby, 519). I choose to believe that Poe's tale is both a warning for egoists who tampered with Nature and also a reflection of what Nature is capable of – since the dull, decaying landscape around the manor is a manifestation of the sick, stagnant family that occupies its premises, and in turn this terrain infects the house and the Usher family. Instead of trying to find a cure to the putrefying topography, Roderick looked inwards into his domestic quarters, trying to separate himself from his surroundings, which eventually destroys him.

Eco-phobia is all about frustrated agency (Estok, 4). Roderick feels castrated because there is nothing he can do to help his family or his mansion from dying. The mirroring of the house Usher and the family Usher, the brother and the sister, Roderick and the narrator, are all set as pieces in a tragedy. To the audience, Usher would mean the family and the mansion as the narrator says, but the true protagonist of Usher is the decaying habitat surrounding the house which is gradually killing its inhabitants. The reason why the characters are always ill in temperament and in bodily misery is a direct symptom of this disease.

The tale begins with a mood of oppressive gloominess; it is as if the world is at a standstill before a storm. The landscape is bleak, desolate, and eerie which the house mirrors. Fear of the loss of agency does strange things to people. Fear of the loss of agency and the loss of predictability are what form the core of eco-phobia (Estok, 4). Roderick's inability to control Nature takes a sinister shape as he tries instead to control the fate of his sister Madeline, whom he buries while she is still alive. Madeline's return from her repressed state in the grave, to fall upon Roderick, subsequently leading to the collapse of the House is then a moment of Nature's triumph over the unnatural (the house and the inhabitants' incestuous history). The Human/Nature and Man/Woman binaries are overturned finally destroying the cancerous mansion and its inhabitants and restoring the natural.

Roderick Usher's tragedy also lies in the fact that he had realized the abnormality of his family and the mansion and had sensed their impending doom. Yet out of sheer egoism and ignorance, he refused to act to mend the injuries inflicted by the damaged House upon Nature. Perhaps this is because he was an organic part of the manor. The more Roderick lost his agency, the more deranged he became. Gradually his identity morphed with that of the house, until both entities became inseparable. The Fall, therefore, much like the Biblical Fall is irreversible and a punishment for the transgression of a natural order by human agency; a triumphant moment when the uncanny in Nature re-emerges with independence.

Conclusion

Therefore, it is pertinent to say that Poe created tragic characters on journeys of self-destruction which is an impulse of humanity at large with our self-centrism. As Estok rightly describes, Human history is a history of controlling the natural environment. Thus, Nature becomes the hateful object in need of our control, the loathed and feared thing that can only result in tragedy if left alone (210). This biased view can only be corrected in Poe with the loss of subjectivity of the human. What is tragic for the characters and the audience is a victory for the natural order. The unnatural in Poe is a civilization with its drive to control. It is this drive that eventually destroys the individual as is evident in both tales. Perhaps the best we can do in eco-critical activism is to let go of the control and to try and view ourselves as part of the ecosystem and not superior to it.

Works Cited

Crosby, Sara L. "Beyond Ecophilia: Edgar Allan Poe and the American Tradition of Ecohorror." Interdisciplinary Studies in Literature and Environment, vol. 21, no. 3, 2014, pp. 513–525. JSTOR www.jstor.org/stable/26430359. Accessed 12 Oct, 2020.

Estok, Simon C. "Ecocriticism in an Age of Terror." *CLCWeb: Comparative Literature and Culture*, vol. 15, no. 1, 2013, doi:10.7771/1481-4374.2182.

Estok, Simon C. "Theorizing in a Space of Ambivalent Openness: Ecocriticism and Ecophobia." *Interdisciplinary Studies in Literature and Environment*, vol. 16, no. 2, 2009, pp. 203–225. doi:10.1093/isle/isp010.

Farrant, Timothy, and Alexandra Urakova. "From 'The Raven' to 'Le Cygne': Birds, Transcendence, and the Uncanny in Poe and Baudelaire." *The Edgar Allan Poe Review*, vol. 15, no. 2, 2014, pp. 156–174. JSTOR www.jstor.org/stable/10.5325/edgallpoerev.15.2.0156. Accessed 12 Oct, 2020.

Freud, Sigmund. "The Uncanny." *The Standard Edition of the Complete Psychological Works of Sigmund Freud*, edited by James Strachey et al., 24 vols, Hogarth, 1955.

Grimshaw, Mike. "The Antimodern Manifesto of the Rural Flaneur: When D'arcy and John Go for a Wander." *The Journal of New Zealand Studies*, vol. 13, 2013, doi:10.26686/jnzs.v0i13.1198.

Hillard, Tom J. ""Deep Into That Darkness Peering": An Essay on Gothic Nature." *Interdisciplinary Studies in Literature and Environment*, vol. 16, no. 4, 2009, pp. 685–695. JSTOR www.jstor.org/stable/44733488. Accessed 12 Oct, 2020.

Husain, Majid. *Human Geography*. Rawat, 1999.

Lefebvre, Henri, and Donald Nicholson-Smith. *The Production of Space*. Blackwell, 2009.

Murphy, Bernice M. *The Rural Gothic in American Popular Culture: Backwoods Horror and Terror in the Wilderness*. Palgrave/Macmillan, 2013.

Poe, Edgar Allan. *The Fall of the House of Usher and Other Tales*. Transaction Publishers, 1998.

Poe, Edgar Allan, et al. *The Raven and Other Poems*. Oxford University Press, 2016.

Punter, David. "The Uncanny." *The Routledge Companion to Gothic*, edited by Catherine Spooner and Emma McEvoy, Routledge, 2007, pp. 129–136.

Rust, Stephen A. and Carter Soles. "Ecohorror Special Cluster: "Living in Fear, Living in Dread, Pretty Soon We'll All Be Dead."" *Interdisciplinary Studies in Literature and Environment*, vol. 21, no. 3, 2014, pp. 509–512. JSTOR, www.jstor.org/stable/26430358.Accessed 12 Oct. 2020.

Part III

Development and Sustainability

14 Development and Sustainability

Understanding the Duality of Expectations through a Study of Literature

Abhisek Bhattacharya and Sudipa Chowdhury

If development in its commonest interpretation implies an introduction of new inventions and institutions to elevate the standards of living, it needs equally to be characterized by sustainability in order that its fruits might bear lasting nourishment to mankind. However, what we find in reality is often the contrary, for, while working upon the measures of growth, we pay but little attention to how the aspects of sustainability are frequently compromised with. This becomes particularly discernible when the processes of urbanization and other geo-economic changes like the construction of river barrages intrude the rural sector and counter-face agrarian economy. It is not that such projects are launched without contemplations of inclusive growth, yet it remains true that the way they encroach upon agrarian land is also consequent upon the residential as well as psychological displacement of a large number of people, often the aboriginal sects, and a loss of their traditional livelihood. In many cases, the calculations of growth thus lead to a paradoxical engendering of issues related to distributional injustice, which, in turn, sharpens criminal propensities among the deprived and depressed mass. These are basically geo-political and economic concerns, but literature cannot stay unresponsive to these issues chiefly because of a writer's obligations to the society. The present chapter is primarily based upon some critical observations on Debesh Roy's (1936–2020) epoch-making Bengali novel *Tista Parer Brittanto*[1] (*Sahitya Akademi* Award, 1990), Charles Dickens's[2] (1812–1870) fictional masterpiece *Oliver Twist*[3] (1839), and finally, Aravind Adiga's Booker-winning Indian English text *The White Tiger*[4] (2008) that might make us cognizant of how the lack of sustainability may render technological and economic growth detrimental to the environmental balance as well as to the notions of distributional justice. The differences of space and time, language, and even ethnicity stand apparent in the choice of the texts themselves, yet, such diversities may serve well for us to draw a sustained conclusion, that is, the fruits of development shall not obtain mobility among the masses until and unless they are warranted by sustainability and the right to parity in distribution.

The chronology of the literary texts we have chosen demands us to start our discussion with Dickens's nineteenth-century tale *Oliver Twist*.

DOI: 10.4324/9781003383970-18

However, our theme prompts us to prioritize the pattern of thought so that the central argument is clearly borne out. For that we feel it more appropriate to begin with Debesh Roy's *Tista Parer Brittanto*, which in every sense is a passionate account of how geo-economic issues might lead to both the physical and psychological estrangement of human beings, mostly the marginalized ones, from their original habitat. Debesh Roy was born in Pabna in British India (now Bangladesh). His family, however, moved to Jalpaiguri in North Bengal in the year 1936. From his days in the college as a student, Roy started working in the student wing of the Communist Party of India. During this period, he learnt the *Rajbanshi* dialect of North Bengal. He also worked as a research fellow in the Centre for Studies in Social Sciences. Roy's life and works were inspired by the Tista river-based Rajbanshi community of North Bengal. The gushing mountain river Tista found its best description and also its lamentable tale of losing natural flow because of geo-economic manoeuvres in Roy's epic-novel *Tista Parer Brittanto*. Rivers have often been the muses of authors all over the world; but the loss of an original riverbed and the plight of the aboriginal communities that thrive and survive along its banks probably found one of the best depictions in Debesh Roy's novel. No wonder he was conferred India's highest literary honour, the Sahitya Akademi Award for the novel in 1990.

The story of the novel centres upon an actual geo-economic event, the construction of a barrage on the river Tista at Gajaldoba in the district of Jalpaiguri in North Bengal (the project was started in 1976, the barrage being inaugurated in 1993). The primary objective of taming down the river's natural flow may be identified as an economic concern for recuperating greater expanses of agrarian land from the formerly flood-prone zones and securing simultaneously a proportionate storage of water to facilitate the cultivation of crops throughout the year. However, it needs to be firmly asserted that while the calculations of an economic growth were so meticulously carried out, the execution of the same mostly stood devoid of proper care towards distributional issues. Hence the possession of the newly obtained agricultural lands was soon monopolized upon by the rich farmers of the region or by wealthy traders from outside. A large number of people in possession of small plots of land suddenly discovered themselves at the face of an unequal contest in the agrarian sector, and very soon they were forced to sell their ancestral land at a cheap price and to embrace the fate of the earners of daily wages. They fell victim to a sort of organized plunder – a plunder not only of land and property, but of dignity, when a man was forced to plough a land, erstwhile his own, under the command of a wealthier outsider.

Debesh Roy's concern for distributional issues attains a greater vitality when he deals with the displacement that the aboriginals like Bagharu and Madari have to suffer because of the geo-economic manoeuvre resultant upon the loss of the original riverbed and the wilderness adjacent to it. The aboriginal people do not know the jargons of geo-economics, what they know instead are the language of the leaves, murmur of the stream, and the whisper of the passing breeze. They are born the children of the forest, having the river as their companion for life. So, when the natural flow of the river is redirected and the forest is destroyed in the name of greater economic benefits, thousands of Bagharus are left traumatized. They frantically search for the river and their lost habitat till they are subdued by a strange sense of defeat. The newly evolved economic zone pulls them not, because they increasingly perceive that they do not belong to it. People with malevolent designs, particularly the outsiders, continue to grasp land after land, while the original inhabitants, the children of the soil, are estranged to measure fathoms of psychological destitution. Debesh Roy's narrative is relentlessly honest in bringing home this duality of expectations. For, a committed reader cannot afford to miss the poignancy of the central image which depicts Bagharu and Madari's antithetical departure from the scene in search of *another forest* adjacent to *another river* when the inauguration of the Tista barrage is to be solemnized in the presence of policy-makers and trade-magnificoes with development pledged to all and sundry. Roy thus concludes, 'With the inauguration of the Teesta Barrage, the Teesta of the story became history. Teesta is no longer the old one.'[5] Here begins the real story of the Tista bank, with the barrage:

> The Teesta Barrage Project sharing water between North Bengal and Bangladesh was later considered to be a huge human catastrophe, as the river was forced into a natural death. In the dry season, only the ice-melted water from Sikkim flows through the river, which is certainly not enough for irrigation ... This water can hardly keep the ecosystem of the basin intact. Both Teesta barrages had been created by spreading unrealistic hopes to a vast population in either country. Frustration is now widespread after disillusionment set in. And the water level in the basin between the two barrages in the two countries is decreasing gradually, which could lead to terrible consequences for the ecosystem in the days to come.[6]

However, Debesh Roy appeals more to the good sense of the readers by portraying the passivity of Bagharu and Madari's response to the central situation. Dickens and Adiga, on the other hand, propel us yet further to be cognizant of the menace of criminal activities which attain undeterred growth at the absence of sustainable development.

Charles Dickens began writing *Oliver Twist* at a time when Jeremy Bentham's[7] Utilitarianism was at full flow in Victorian England. The necessity of economic rejuvenation was felt to recover the vast expenses of industrial production. But this also led the English people, particularly the business magnates, to value almost everything in terms of its practical utility. Enactment of the Utilitarian economy soon resulted in a harvest of wealth that was primarily meant to ensure the financial progress of the nation. But we should remember that the Utilitarian preference for urbanization also led to the collapse of rural economy. The sweet breath of country air was contaminated by filthy emissions of the newly built factories. A number of people formerly enjoying economic independence in the agrarian sector suddenly found themselves in the streets because of large-scale urbanization. They were bereft not only of home but also of their traditional livelihood. Fertile lands were transformed into heaps of concrete and iron. Countryfolk were led to accept enforced slavery in factories and mills of the towns for mere sustenance. Apparently, the picture of urban development attained prominence with a panorama of multi-storeyed and multihued office buildings and residencies, but at the same time, nasty, unhygienic, and clumsy slums that were built at the outskirts of the towns demonstrated a hideous contrast to what was expected to bring comprehensive development in the society and economy of nineteenth-century England. On the one hand, the business tycoons continued to heap wealth, while on the other hand, life of a large number of people, who were deracinated from their traditional livelihood, was drowned in a widening abyss of economic as well as social destitution. Distributional injustice robbed them off the basic rights of living. Thwarted expectations soon drove them to criminal acts in order to secure the sustenance of their loved ones. Development that Utilitarian economy had promised thus failed to sustain, and consequently, an obnoxious system that was characterized by robbery and murder came to parallel the systematic forgery of the privileged Londoners.

The novel *Oliver Twist* holds perhaps the most prominent illustration of the London underworld in the realm of nineteenth-century British fiction. The novelist forgets nothing. On the one hand, he shows opulence of wealth under the possession of a small number of business tycoons. Then, as a contrast to their ambivalent economic superiority, he inserts into the narrative contemporary issues of spoilt childhood and limitless exploitation of child labour, which were responsible for the moral derangement of adolescent boys and girls. Oliver Twist, the protagonist of the eponymous novel, is born in a workhouse in 1830s England. His mother, whose name no one knows, is found on the street and dies just after Oliver's birth. Oliver spends the first nine years of his life in a badly run home for young orphans and then is transferred to a workhouse for adults. Oliver, starved and exhausted in his later days in London, meets Jack Dawkins, a boy of his own age. Jack offers him shelter in the London house of his benefactor, Fagin. It turns out that Fagin is a career criminal who trains orphan boys to pick pockets for him. Not only Oliver, but in the present novel, we see a number of juvenile delinquents of both sexes by the side of monstrous miscreants. This bears sufficient evidence of the state's failure in doing justice with distributional issues. The optimist in Dickens finally relieves the gloom by showing the recuperation of Oliver at the dainty of the benevolent Mr Brownlow. We shudder still to think of countless adolescents of his age and disposition, whose expectations in life were totally shattered because of the forgery of the rich. Policymakers prescribed theoretical remedies and economists spent hours in calculations, but there was taken hardly any practical measure that could make development sustainable by levelling distributional disparities. Fagin was hanged at the end, but it is also understandable that hanging one Fagin does not put any end to criminal activities. Criminal acts are certain to recur until and unless the breach between development and sustainability could be bridged.

Aravind Adiga is the third writer of our choice for the present review, and his locale is twenty-first century India. What we find in the narrative of *The White Tiger* is an unmistakable class-consciousness and an urge among the under-privileged Indians to secure for themselves equal rights of living erstwhile monopolized upon by their privileged counterparts. Interestingly, the group of under-privileged people in the novel is shown to be peculiarly unabashed even if they have to choose the most derogatory of means to succeed in their mission. Thus, it becomes once more perceivable to us how the lack of sustainable development serves to whet criminal propensities. In this connection, we should remember Prof. Amartya Kumar Sen and Prof. Jean Dreze's opinion[8] that in India, possibilities of sustainable development are shattered chiefly because of distributional injustice. They termed the same as a sort of *disguised violence* done to the under-privileged mass by a handful of rich people. In Adiga's *The White Tiger* we come to perceive the truth of Prof. Sen and Prof. Dreze's observation. Balram Halwai, a rickshaw puller's son from an impoverished background, is born in the village of Laxmangarh. In Laxmangarh, Balram was raised in a large, poor family from the Halwai caste, a caste that indicates sweet-makers. The village is dominated and oppressed by four landlords. Balram was initially referred to simply as 'Munna,' meaning 'boy,' since his family had not bothered to name him. He did not have another name until his school-teacher dubbed him Balram. The boy proved himself intelligent and talented and was praised one day as a rare 'White Tiger' by a visiting school inspector. Unfortunately, Balram was removed from school after only a few years, to work in a tea shop with his brother, Kishan. There, he furthered his education by eavesdropping on the conversations of the shop customers.

Balram feels that there are two Indias: the impoverished one darkened by the ill-fated poor, and the rich other, illuminated by a thousand lustres of the wealthy business magnates. However, he strongly holds that poverty involves both the deliberate methods used by the upper class and a servile mentality enforced upon themselves by the undercasts. Balram's father died of tuberculosis in a decrepit village hospital, where no doctors were present due to abundant corruption within all the government institutions. After their father's death, Kishan got married and moved with Balram to the city of Dhanbad to work. There, Balram decided to become a chauffeur and took driving lessons from a taxi driver.

As a driver in his later days in the household of Mr Ashok, Balram lived a stable and satisfactory life. He describes at length the corrupt nature of politics in the country. His master's family, involved in shady business dealings in the coal industry, must regularly bribe the politicians to ensure their success. Eventually Balram also came to learn how his masters kept illegal connections with the police by means of their monetary power.

In the course of time, Balram's political consciousness grows more intense, and his resentment towards the upper class more violent. Much of the narrative then traces his growth from a meek peasant to an inflamed individual capable of murder in pursuit of his own success.

Finally deciding to break free from the forgery of the master class, Balram fashioned a weapon from a broken whisky bottle and rammed the bottle into Ashok's skull, and then stabbed him in the neck, killing him. He stole the master's red bag, filled with a handsome amount of money and escaped to Bangalore. In Bangalore, Balram found great success. He launched a taxi service for call centre workers, which he calls 'White Tiger Technology Drivers.' By bribing the police, Balram was able to gain influence and make his business successful. Demonstrating how far he has come, he refers to how he has succeeded in hiding a fatal accident through his connection with the authorities. He considers his life to be a quintessential entrepreneurial success story. What is more, in his frank confession, he continues to generalize his method as one of the standard methods of becoming rich in India. He exalts on his success, but his almost devilish exaltation comes to evoke our shame and apprehension. We are left speechless in apprehension about the future of India in the forthcoming decades under the impact of distributional injustice.

Whether in Dickens or in Adiga, the picture therefore remains more or less the same. While Debesh Roy ended his narrative by depicting the passivity of human responses in *Tista Parer Brittanto*, Dickens in *Oliver Twist* tended to be optimistic. In his days men might have been left with hope. But, after almost two centuries, when Adiga came to write his first English-language novel, he found not the least streak of hope anywhere. The difference of time and place seems to matter little; what really comes to strike us is the steady denigration of human values under the mantra of 'development,' the lack of sustainability of which is due to its failure in doing justice to the basic requirements of ordinary humanity. One thus cannot but admit that man's psychological estrangement resultant upon criminal activities will never come to an end, should the fruits of development remain forever beyond the reach of people, who are marginalized and as such, belong to the lowest and financially lame stratum of the human society.

However, a litterateur is never a reformer, nor is his work comparable to a socio-economic or even political treatise based on a collection and analysis of objectively valid data. He depends more upon a direct acquaintance with the lived experiences of men and allows his creative imagination to build upon the primary impressions obtained during his interface with the other men and women. A literary interpretation of any crisis of humanity is, therefore, essentially subjective in nature, unlike the objective analyses of the social sciences, and it continues to appeal more to the emotions of the readers and the audience than to intellect. Intellectual analyses may have an archival importance, but the interpretations of art and literature are more dynamic, as they can generate and motivate deep-rooted psychological responses among general people. These psychological responses are too deep seated to suffer wane with the passage of time, whereas the objective data of Sociology and Economics often suffer the confines of a temporal and spatial exclusiveness. In the present chapter we have made a conscious choice of three different texts from three different literary and cultural milieus. Debesh Roy's *Tista Parer Brittanto* is a *Bhasa* text written in the late 1980s, while Aravind Adiga's *The White Tiger* is a twenty-first century Indian English novel, and Charles Dickens's *Oliver Twist* an English classic of Victorian London. The difference of time and perspective in the texts is obvious, but what impels us to bring them into a single frame of study is the concern for man's persistent plight at the absence of sustainable development, and this concern for justice in distributional issues has been shared in common by each one of the writers despite their chronological as well as racial differences. Dickens therefore still continues to be a contemporary of Debesh Roy and Adiga, and this sense of contemporaneity valorizes the relevance of the literary impression in all ages that are to follow. The litterateurs, as we have already noted, are no social or political or even economic reformers to articulate any definite remedy or solution of the crisis. Their art is finer than the methodology followed in social tracts or economic treatise. They go on registering the subdued feelings of the oppressed and their hushed-up tales of misery in the perennial lines of literature. Dickens's Oliver therefore finds another face in the thousand Bagharus and Madaris of North Bengal. Similarly, Adiga's Balram Halwai comes to represent generations of ill-fed and ill-treated people all over the world, whose lived experiences of bitterness lead them to embrace the darkness of the underworld activities. What the government would decide for a comprehensive well-being of these people might be debated by the statesmen and the economists. The rehabilitation projects subsequently initiated might even introduce newer gateways for later scholars to theorize upon. But, for the present at least, the existing literature of the world, beyond the frontiers of every possible jargonistic selectiveness, is expected to hold before us a looking glass, whereon we shall gaze to find the face of an Oliver or a Balram or a Bagharu in the face of each one of us. This is also expected to lead us to achieve a cathartic effect, rather than to agitate ourselves by a thousand vexations of socio-political or even economic objectivity.

It is, however, a matter of relief and assurance that the Government of India, like many other civilized governments of the world, has initiated a number of rehabilitation projects for the deprived people over the last few decades, casting off what may be termed a prolonged hibernation. Among these projects, the *Pradhan Mantri Jan Dhan Yojana* (PMJDY) is a national mission for financial inclusion to ensure the access of an unbanked Indian citizen to financial services like basic savings and deposit accounts, remittance, credit, insurance, and pension in an affordable manner. Under the scheme a basic savings bank deposit (BSBD) account can be opened in any bank branch or business correspondent (Bank Mitra) outlet within the country by persons not having any other similar account. There is no need of maintaining any minimum balance in a PMJDY account.

Another such governmental project is the *Pradhan Mantri Awas Yojana*, which is an initiative to provide affordable housing to the urban as well as the rural poor. The target is to build 20 million of such houses by 31 March 2022. The project was launched on 25 June 2015. Besides, the Government of India also launched the *Deen Dayal Upadhyaya Gram Jyoti Yojana* to provide continuous electricity to rural India in the year 2015. The government plans to invest 756 billion rupees for rural electrification under this scheme. The scheme replaced the erstwhile *Rajiv Gandhi Grameen Vidyutikaran Yojana* in 2015.

The PMUY or the *Pradhan Mantri Ujjwala Yojana* launched by the Government of India in 2016 aims to distribute 50 million LPG connections to the women of BPL (below poverty line) families of the country. Another significant move in the direction of women empowerment is the *Beti Bachao Beti Padhao* campaign of the Government of India that aims to generate awareness and improve the efficiency of welfare services intended for girls in India. The scheme was launched in 2015 with an initial funding of rupees 100 crore (14 million). It mainly targets the socially and economically backward clusters of people in Uttar Pradesh, Uttarakhand, Haryana, Punjab, Bihar, and Delhi.

Whether these governmental initiatives will succeed in levelling economic disparities in India and among its people who suffer displacement because of geo-economic changes, might adequately be rehabilitated in future, may well be debated in the academy. But the literary sensibility is not to be satisfied by promises or controversies. It receives sustenance from the actual lived experiences of men and pleads to the good sense of men again for a restoration of justice and order in a world molested by ravenous greed and shameless Mammonism. The crusade for the basic rights and conditions of living will therefore continue in the world irrespective of spatial or racial differences. Let the sociologists and the economists of the world bring about fruitful research for the resolution of the ongoing crisis of civilization. The literary sensibility shall be ever awake at the interval to greet a new dawn that will appear more human and humane, having the forces of discrimination diminished in or exiled from the annals of humankind. Viewed from this greater perspective, Dickens, Roy, and Adiga shall remain the staunch crusaders for justice despite the different time and clime to which they belonged, to hold the humanistic ensign beyond every yardstick of time, space, and culture-specific constraints that we often apply as invincible fixations in the understanding and appreciation of literature and the fine arts.

Notes

1 Debesh Roy was born in 1936. He planned *Tista Parer Brittanto* on the displacement of a large number of people during the construction of the Tista Barrage at Gajaldoba in the district of Jalpaiguri in the 1980s. The novel was first published in 1988, and it won the Sahitya Akademi Award in 1990.
2 Charles John Huffam Dickens (7 February 1812–9 June 1870) was an English writer and social critic. He created some of the world's best-known fictional characters and is regarded by many as the greatest novelist of the Victorian era. See https://en.wikipedia.org/wiki/Charles_Dickens.
3 *Oliver Twist; or, the Parish Boy's Progress* is Charles Dickens's second novel and was first published as a serial from 1837 to 1839. The story centres on orphan Oliver Twist, born in a workhouse and sold into apprenticeship with an undertaker. After escaping, Oliver travels to London, where he meets the 'Artful Dodger,' a member of a gang of juvenile pickpockets led by the elderly criminal, Fagin. See https://en.wikipedia.org/wiki/Oliver_Twist.
4 Adiga, Aravind. *The White Tiger*. HarperCollins, 2008.
5 https://www.getbengal.com/details/story-of-teesta-died-long-ago-and-now-debesh-roy-who-brought-teesta-alive-dies
6 Ibid.
7 Jeremy Bentham (15 February 1748–6 June 1832) was an English philosopher, jurist, and social reformer regarded as the founder of modern utilitarianism. See https://en.wikipedia.org/wiki/Jeremy_Bentham
8 See Dreze, Jean and Sen, Amartya. *India: Development and Participation*. Oxford, 2002. pp. 222–223.

Works Cited

Adiga, Aravind. *The White Tiger*. New York: Atlantic Books, 2008.
Dickens, Charles. *Oliver Twist*.India: Fingerprint Publishing, 2016.
Dreze, Jean and Sen, Amartya. *India: Development and Participation*. London: Oxford University Press, 2002.
Roy, Debesh. *Tistaparer Brittanto*. Kolkata: Dey's Publishing, 1988.

15 Sustainable Development and Ecological Perspectives

Improvement in Water and Sanitation

Chandan Bandyopadhyay

Water is essential for human civilization. Without water no living being can survive; so a sufficient amount of water is necessary for human beings. Access to water is considered a right. International bodies confirm this right in their agenda, meetings, treaties, declarations, etc. Sanitation is also considered as essential as food and water among other things for living a standard life. Both water and sanitation can make our life hygienic, free from diseases, and healthy.

A wide range of ecological and human crises result from inadequate access to, and the inappropriate management of, freshwater resources. As human population continues to grow, these problems are likely to become more frequent and serious. New approaches to long-term water planning and management that incorporate principles of sustainability and equity are required and are now being explored by national and international water experts and organizations (Gleick, 1998). Environmental degradation caused due to civilization makes our life more vulnerable. High population growth, urbanization, and rapid industrialization have indeed polluted our environment which has resulted in climate changes. Agricultural activities are also affected negatively due to environmental pollution and climate change, thereby resulting in food insecurity also. Availability of water for agricultural and industrial purposes along with domestic needs is scarce due to the high use of groundwater in absence of rainwater and also due to the use of unscientific production techniques.

Safe drinking water and sanitation are now as fundamental factors as food, shelter, etc. The absence of water and sanitation may lead to health problems mainly to the vulnerable sections like children and women, aged persons, diseased people, and also socially and economically vulnerable people. This may affect enrolment rate, specifically girls' rate of enrolment. There may be low productivity and high morbidity due to the lack of water and sanitation. It also restricts women's option of alternative employment opportunities as they have to engage some extra hours in securing water from a distance for their families and also by serving children and elderly people of the family ailing due to contamination of water and unavailability of sanitation facility.

DOI: 10.4324/9781003383970-19

Universally, women usually serve as domestic water managers, concerned with both domestic water provision and use. This task of women, particularly in the rural areas, has been constrained by factors like inaccessibility and non-reliability of water sources, arising primarily from problems with water quantity and quality. Women have to bear the burden of the family in the context of water and sanitation. For example, if safe drinking water is not available within the premises or nearby then it has to be collected and fetched from a distance. UNICEF said the 200 million hours which women and girls spend every day globally in collecting water is a colossal waste of their valuable time (UNICEF, n.d.). Despite investments from various corners, only 28% of people in rural areas in India have access to toilets that leads to a severe burden on the community, health services, and considerable losses in productivity. Again likelihood of being affected by several water-borne diseases like diarrhoea, cholera, typhoid, and malaria increases due to the absence of safe water and sanitation (Purohit, 2012). Children and elderly persons in the family are mainly affected by these diseases. Women take care of them and thereby spend extra hours, thus limiting their time in search of alternative employment opportunities.

Nowadays in our world, cities are continuously becoming water-scarce regions to support their ever-increasing population and related activities due to the near exhaustion of water bodies. Economies with large population and heavily dependent on agriculture faced challenges to support their ever-increasing demand of water. Water management policies may somehow mitigate the problem. However, typical water development project is undesirable from a sustainability perspective, as it is usually associated with serious ecological and social impacts as well as suboptimal cost-effectiveness. For this investment for developing infrastructural facility, public–private partnership model and urban–rural partnerships are necessary in this modern developed world (Richter et al., 2013). A large number of households in cities around the developing world also do not have access to a safe and reliable supply of drinking water (McKenzie & Ray, 2009). During the 1980s and 1990s, efforts had been made through several plans to develop infrastructure for providing safe water and sanitation facilities to the common people in India. However, the current system of supplying water in the household for a large section of the population is inadequate (McKenzie & Ray, n.d.).

Goal 7 of the Millennium Development Goal (MDG) was to ensure environmental sustainability and within this Goal target, 7C was to halve the proportion of the population without sustainable access to safe drinking water facilities and basic sanitation. After assessing the MDGs in 2015 the Sustainable Development Goals (SDG) were framed with the provision of achieving these goals within 2030. Among the goals, Goal 6 is to ensure availability and sustainable management of water and sanitation for all which is more targeted than the MDG target to halve the proportion of the population without access to water and sanitation by 2015.

Poverty eradication and ensuring decent work by achieving sufficient growth are also framed as goals in both MDG and SDG; Goal 4 of the SDG is to ensure inclusive and equitable quality education for all, and among the targets within this goal ensuring all types of quality education to girls and women without discrimination is to be achieved within 2030. At the same time, a consensus is growing worldwide that water and water services are essential because they touch on almost all Millennium Development Goals. Investment in water infrastructure – to protect against droughts and floods, produce renewable energy, and provide water supply to cities and rural areas, and water to grow food – is basic for economic growth and poverty reduction in poor countries.

A specific challenge is the fact that a growing world population requires more water services and water infrastructure for the underlying economic activities that enable development, notably to provide for safe water supply and sanitation, meet increasing food needs, and secure inputs for industrial processes. At the same time, however, our ecosystems, which provide valuable services to us and sustain life on Earth, also need sufficient amounts of water to function. For this reason, sustainable development *and* management of water is a must in the responsible-growth equation. Thus, we have seen that there are targets like poverty reduction, providing education to all, and also ensuring safe drinking water and sanitation facility in the SDGs. The globalized order is now experiencing both the rise of poverty and the local and global consequences of water scarcity, and thus the issue of finding the relationship between water and poverty is the object of sprawling literature (Molle & Mollinga, 2003).

United Nations General Assembly took a resolution regarding water and sanitation as fundamental human rights (UN General Assembly, 2010). The Convention on the Elimination of All Forms of Discrimination against Women (CEDAW) obliges governments to ensure that rural women have access to sanitation. The Convention on the Rights of the Child (CRC) recognizes the right of all children to an adequate standard of living. The UN expert body responsible for monitoring the CRC has clarified that this entitlement includes access to clean drinking water and latrines. The UN Human Rights Council adopted a resolution in March 2008 emphasizing that international human rights law, including the International Covenant on Economic, Social and Cultural Rights (ICESCR), CEDAW, and CRC, entails obligations concerning access to sanitation. The 1949 Geneva Conventions entitle prisoners of war and civilian internees to access sanitation in situations of armed conflict and occupation. As is stated by the UN, 'the human right to water entitles everyone to sufficient, safe, acceptable, physically accessible and affordable water for personal and domestic use,' where a sufficient quantity of water should correspond to WHO guidelines. Although it is difficult to set a universal water poverty line due to regional differences in climate and basic needs, a minimum of 20 litres a day from a source within one kilometre of the household has been suggested. This is the level of water that is sufficient for drinking and basic personal hygiene (WHO and UNICEF, 2000).

Here are some studies which specifically show the importance of water and sanitation in urban planning, controlling diseases, and other developmental aspects. Amitabh Kundu in his paper "Micro Environment in Urban Planning: Access of Poor to Water Supply and Sanitation" (Kundu, 1991) attempted to examine the nature and magnitude of disparity in the access to water supply and sanitation of people in different levels of consumption expenditure in urban areas. He showed that the public water supply system is heavily subsidized by government grants. It is also shown that there is a lesser percentage of people lying below the poverty line being supplied piped water as compared to the overall total population. This is true for the service of sanitation facilities as well.

V. Ratna Reddy in his paper "Declining Social Consumption (Drinking Water and Sanitation) in India" (Reddy, 2001) showed that there is a declining trend in the availability of drinking water even though drinking water and sanitation have always been on top of the priority list of social consumption items. The main reason is the lack of efficiency in the policymaking and also in the implementing process. Johanna Modigsson and Ylva Munkhammar in their thesis "The Dimensions of Water Scarcity: A Study of Development Priorities in Sub-Saharan Africa" (Modigsson & Munkhammar, n.d.) explained the relationship between water availability and economic as well as physical factors in 45 countries in sub-Saharan Africa. Total water coverage is considered a dependent variable. Total actual renewable water resources per capita are treated as a physical factor. Gross domestic product per capita, Official Development Assistance to water per capita, annual population growth rate, Civil Liberties Index, and female percentage of secondary education enrolment are treated as economic factors. They showed that GDP and female enrolment have a strong positive relationship with the dependent variable compared to population growth and civil index.

Rashmi Tiwari and Sanatan Nayak in their paper "Drinking Water, Sanitation and Waterborne Diseases" (Tiwari & Nayak, 2015) have examined the relationship between water, sanitation, and waterborne diseases using the data of two districts (Lucknow and Kanpur) of Uttar Pradesh. They showed that the factors like caste, education, and income determine access to safe and clean drinking water. It has also been shown that a good source of drinking water has a significant impact on human health. Gyana Ranjan Panda and Trisha Agarwala in their paper "Public Provisioning in Water and Sanitation: Study of Urban Slums in Delhi" (Panda & Agarwala, 2013) examined the delivery of essential services like drinking water and sanitation to the people living in slums as a policy priority for the government. Judging from the Eleventh Plan Budgetary allocation to the water supply and sanitation services they showed that the budgetary outlay for slums is uncharacteristically inadequate.

Gomez, Perdiguero and Sanz in their paper analysed access to 'total improved,' piped on premises, as well as other improved sources of access in rural areas for low-income, low-middle-income, and high-middle-income countries. They suggested that gross national income (GNI); female primary completion rate; agriculture; growth of rural population; and governance indicators, such as political stability, control of corruption, or regulatory quality, are variables related to water access and the policy makers (Gomez et al., 2019).

Several literatures established that the well-developed water and sanitation infrastructure, such as tapped drinking water, availability of water within premises, latrine facility, can be availed by those households having good income level, female enrolment rate, etc. However, this infrastructure can also explain or affect the development of an economy.

We know that an economy comprises of mainly three sectors: agriculture, industry, and service. Agricultural activities require a huge amount of water; industrial units also require a good amount of water, particularly in cooling their furnaces; services like constructions, tourism, hotels also require water for different purposes. Thus, water plays as an input (direct) in these production processes. Again, in the modern world labour is termed as human capital (Mincer, 1993) whose productivity increases with their education level and health condition. A safe drinking water and good hygienic sanitation facility improves the health condition of human labour or at least prevents to affect from several waterborne diseases, thereby keeping intact the human capital. In this way it can be treated as an input (indirect) also. In either case, this input has a positive impact on the economic activity which improves the development of the economy as a whole. There may also be an environmental impact of larger economic activity. The Kyoto Protocol, 1997, categorically pointed out that the continuous increase in emission of Green House Gases (GHGs) with the modern industrialized units causes global warming which have effects like environmental degradation and ecological imbalances. The above relationship is explained with the help of the following flow chart.

The following flow chart contains four sectors: Economy, Human Capital, Water and Sanitation, and Environment. The arrows with +ve sign portray positive impact or influences while those with –ve sign portray negative impact or influences. Let us try to explore the idea behind the drawing of this flow chart. Proper development of water and sanitation infrastructure prevents several waterborne diseases to occur which has a positive impact on human capital, increased productivity of human labour enhances economic activities which may increase emission of GHGs and thereby have negative impact on environment. Sufficient water supply increases the economic activity through increase in the production of agricultural products, industrial outputs, and also uninterrupted activities in the service sector. However, unscientific and excessive usage of ground water may result in decrease of water level, increase the arsenic level in the available ground water and thus degrades environment. On the other hand, good environment improves the water quality as well as the capacity of human capital. Again, developed economy can make good education and health infrastructure which may improve the human capital through knowhow and better health. Developed economy also can build better water and sanitation infrastructure. Thus, we have positive as well as negative arrows coming from and going to four sectors in this chart explaining as well as portraying the relationship among them. The World Bank estimated the economic impacts of inadequate sanitation in India in the year 2006 – showing an annual economic impact of Rs. 2.4 trillion (US$ 53 billion), implying a per capita annual loss of Rs. 2,180 (US$ 48) or 6.4% of the GDP in the same year (World Bank, 2011) (Figure 15.1).

This above flow chart contains four sectors: *Economy, Human Capital, Water and Sanitation, and Environment.* The arrows with +ve sign portray positive impact or influences while those with –ve sign portray negative impact or influences.

Source: Author's own drawing.

In Table 15.1, the percentage of households having access to safe drinking water and access to improved latrine facility within premises in rural, urban, and in the states as a whole is shown. From the table it is clear that in several states (13 out of 31, in red) there remains more than 20% households not having access to safe drinking water; this picture fades further in the rural region of states (16 out of 31, in red). The table also shows that in 11 states more than 60% of the households are not having access to improved latrine facility within the premises while in the rural region of those states 75%

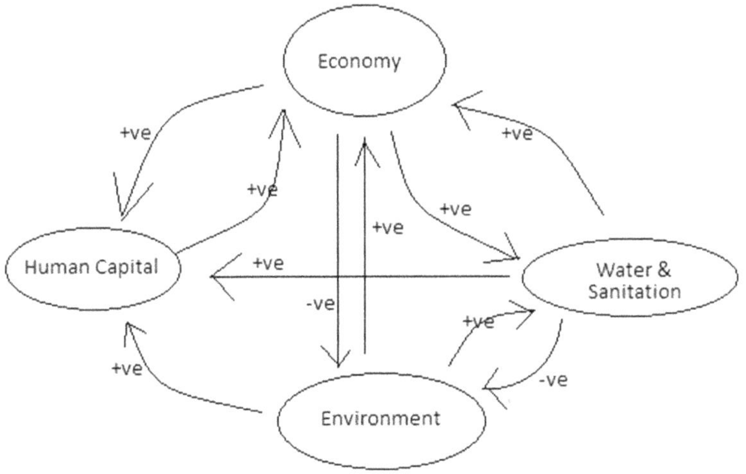

Figure 15.1 Relating Economy, Human Capital, Water and Sanitation, and Environment.

Table 15.1 Relating Economy, Human Capital, Water and Sanitation, and Environment

States	*Access to Safe Water (Tap/ Handpump*		*Drinking*/ Tubewell) (%)*	*Access to Improved Latrine within Premises (%)***			*Contamination-wise ***Number of Habitations*		*DALY Rate Attributed to Unsafe water, sanitation and handwashing*****
	Total	*Rural*	*Urban*	*Total*	*Rural*	*Urban*	*Arsenic*	*Fluoride*	
Andaman and Nicobar	85.5	78.2	98.1	NA	NA	NA	0	0	NA
Andhra Pradesh	90.5	88.6	94.5	46.9	30.3	82	0	349	1238
Arunachal Pradesh	78.6	74.3	91.3	32.8	19.3	72.7	0	0	1099
Assam	69.9	68.3	78.2	30.6	21.8	77	4518	285	2285
Bihar	94.0	93.9	94.7	19.5	14.3	63.2	871	898	2928
Chhattisgarh	86.3	84.1	93.9	21.2	10.7	58.6	19	405	2224
Goa	85.7	78.4	90.4	74.5	64.5	80.8	0	0	272
Gujarat	90.3	84.9	97.0	56	31.4	86.6	0	0	879
Haryana	93.8	92.0	96.7	61.8	48	85.4	0	119	981
Himachal Pradesh	93.7	93.2	97.8	66.1	63.5	86.7	0	0	678
Jammu and Kashmir	76.8	70.1	96.1	31	18.8	66.3	0	4	787
Jharkhand	60.1	54.3	78.4	20.5	6.3	64.6	102	552	3332
Karnataka	87.5	84.4	92.3	48.9	26.5	81.9	4	600	1149
Kerala	33.5	28.3	39.4	89.9	87.5	92.4	0	34	363
Madhya Pradesh	78.0	73.1	92.1	26.5	10.9	71.5	0	171	1879
Maharashtra	83.4	73.2	95.7	50.2	34.8	68.6	0	81	835
Manipur	45.4	37.5	60.8	46.4	37.7	63.8	0	0	1287
Meghalaya	44.7	35.1	79.5	36.4	23.4	84.2	0	0	1294
Mizoram	60.4	43.4	75.8	69.2	51.1	85.6	0	0	848
Nagaland	53.8	54.6	51.8	48.9	37	78.4	0	0	510
Odisha	75.3	74.4	79.8	18.2	10.6	59.6	0	105	2904
Puducherry	97.8	99.6	97.0	NA	NA	NA	0	0	NA

Punjab	97.6	96.7	98.9	72	61	89.4	720	304	793
Rajasthan	78.1	72.8	94.3	29.8	14.9	75.5	0	6159	1570
Sikkim	85.3	82.7	92.2	78.1	72.8	91.9	0	0	478
Tamil Nadu	92.5	92.2	92.9	45.8	21.5	71.9	0	0	862
Telangana	NA	NA	NA	NA	NA	NA	0	987	1330
Tripura	67.5	58.1	91.9	62.5	55.1	81.5	0	0	1478
Uttar Pradesh	95.1	94.3	97.9	31.4	17.9	77.6	748	179	2421
Uttarakhand	92.2	89.5	98.7	63.1	51.4	91	0	0	970
West Bengal	92.2	91.4	93.9	48.5	33.8	80.5	10928	1336	1010
India	85.5	82.7	91.4	41.7	25.1	77.3	17910	12568	1626

*Source: Economic Survey, 2013–14; Office of the Registrar General, Ministry of Home Affairs.
**Source: House listing and Housing Census Data Highlights – 2011; Houses, Household Amenities and Assets- Census of India, 2011.
***Annexure referred to in reply to part (a) and (b) of Unstarred Question No. 1626 to be answered in Lok Sabha on 28.12.2017 regarding 'Ground Water Contamination' State/UT-wise number of fluoride, arsenic, iron, nitrate, salinity, and heavy metals affected habitations, as reported by the States into Integrated Management Information System (IMIS) of the Ministry as on 26 December 2017.
****Source: India: Health of the Nation's States.
Significantly lower than national mean
Indistinguishable from national mean
Significantly higher than national mean
Note: Disability-adjusted life years (DALYs) is defined as years of healthy life lost to premature death and suffering. DALYs are the sum of years of life lost and years lived with disability.

of the households do not have improved latrine facility. Contamination of ground water by arsenic and fluoride is also observed in several states at an alarming rate. This lack of water and sanitation infrastructure surely affects the inhabitants of those states by infecting them with several diseases.

Indian Council of Medical Research (ICMR) in association with Public Health Foundation of India and Institute for Health Metrics and Evaluation made a report on "Disease Burden Trends in the States of India: 1990 to 2016." In this report it was observed that unsafe water and sanitation was the second leading risk responsible for disease burden in India in 1990, but dropped to the seventh leading risk in 2016, contributing 5% of the total disease burden, mainly through diarrheal diseases and other infections. The burden due to this risk is higher in females than in males. Remarkably, the per person disease burden due to unsafe water and sanitation was 40 times higher in India than in China in 2016. The massive effort of the ongoing Swachh Bharat Abhiyan in India has the potential to improve this situation (India: Health of the Nation's States).

Thus, water and sanitation play significant roles in human life. It is the right of every human being to have safe drinking water and adequate sanitation facility to live a secure healthy life. The states should have policies to facilitate its citizens with safe drinking water and adequate sanitation. The economic activities of agriculture and industry also require sufficient availability of water. Proper usage of water, especially ground water, is important, otherwise water level will be depleted and there will be more contamination with arsenic as well. Environmental degradation and climate change also affect our lives. We have to upgrade our production methods with newer environment-friendly technologies to negate the bad impacts. The development goals will be achieved within the stipulated time if the states build infrastructural facilities for providing water and sanitation to their citizens and for this sufficient budgetary allocation is necessary. The ongoing Swachh Bharat Abhiyan in India has the potential to facilitate hygienic sanitation. This will improve the health condition and decrease the burden of death due to diarrhoea.

Works Cited

Gleick, P.H. "Water in Crisis: Paths to Sustainable Water Use." *Ecological Applications*. 8.3 (1998): 571–579. https://doi.org/10.1890/1051-0761(1998)008[0571:WICPTS]2.0.CO;2.

Gomez, Mabel, Jordi Perdiguero, and Alex Sanz. "Socioeconomic Factors Affecting Water Access in Rural Areas of Low and Middle Income Countries." *Water*. 11.2 (February 2019): 202. https://doi.org/10.3390/w11020202.

Kundu, A. "Micro Environment in Urban Planning: Access of Poor to Water Supply and Sanitation." *Economic and Political Weekly*. 26.37 (1991): 2167–2171. JSTOR.https://www.jstor.org/stable/41627002.

McKenzie, D. and Ray, I. *Household Water Delivery Options in Urban and Rural India*, n.d. 59.

McKenzie, D. and Ray, I. "Urban Water Supply in India: Status, Reform Options, and Possible Lessons." *Water Policy*, 11.4 (2009): 442–460. https://doi.org/10.2166/wp.2009.056.

Mincer, J *Studies in Human Capital. 1. "Investment in Human Capital and Personal Income Distribution"*. Edward Elgar Publishing, 1993.

Modigsson, J. and Munkhammar, Y. UPPSATSER.SE: "The Dimensions of Water Scarcity. A Study of Development Priorities in Sub-Saharan Africa". Retrieved July 21, 2020, n.d. https://www.uppsatser.se/uppsats/a0ad711ae7/

Molle, F. and Mollinga, P. "Water Poverty Indicators: Conceptual Problems and Policy Issues". *Water Policy*. 5.5–6 (2003): 529–544. https://doi.org/10.2166/wp.2003.0034.

Panda, G. and Agarwala, T. "Public Provisioning in Water and Sanitation: Study of Urban Slums in Delhi." 48 (2013): 24–28.

Purohit, B.C. "Health Impact of Water Borne Diseases and Regional Disparities in India." *International Journal of Health Sciences*. 1.18 (2012). 122–131

Reddy, V. "Declining Social Consumption (Drinking Water and Sanitation) in India." *Economic and Political Weekly*. XXXVI (2000). 192–204.

Richter, Brian D., David Abell, Emily Bacha, Kate Brauman, Stavros Calos, Alex Cohn, Carlos Disla, et al. "Tapped out: How Can Cities Secure Their Water Future?" *Water Policy* 15.3 (June 2013): 335–363. https://doi.org/10.2166/wp.2013.105.

World Bank. *New Delhi. Water and Sanitation Program "Economic Impacts of Inadequate Sanitation in India."* World Bank, 2011.

16 Analysis of Ambient Air Quality of Asansol Subdivision and Its Sustainable Solution

Sarbendu Bikash Dhar

Introduction

Good air quality is one of the basic criteria of human well-being. It has been found that a person inhales more or less 14,000 litres of air daily. The presence of pollutants in the air can adversely affect people's health as well as the natural environment. Pollutants can also settle out of the air onto the earth's surface features like soil, water bodies, and leaves. From the topsoil layer, those pollutants may leach into the groundwater or may be taken up by plants and animals. Adverse air quality can also affect the climate (EEA, 2013).

Probably every developmental activity is associated with some kind of pollution effect. Thus, pollution is an unavoidable evil of development. Different kinds of human actions release harmful compounds; the burning inputs in the atmosphere are oxides of nitrogen (NO_2), sulphur dioxide (SO_2), and tiny solid particles called particulate matter (PM_{10}, $PM_{2.5}$) (Goyal & Sidharth, 2003).

All these air pollutants have harmful effects on human health and the environment. Pollutants from these sources may not only prove a threat near these sources but can travel extensive distances, chemically reacting in the atmosphere to produce acid rain or ozone holes. Petrol and diesel engines from motor vehicles emit a wide variety of pollutants, principally oxides of nitrogen (NO_2) which have an increasing impact on air quality, especially in urban regions (Mage et al., 1996).

Nowadays, air pollution is a serious problem almost in every urban centre. Out of the 34 most polluted cities across the world, 21 are located in India. It has been found that 51% of pollution is caused by industries, 27% by vehicles, 17% by crop burning, and 5% by Diwali fireworks. In urban areas, vehicular emissions and industries are the major sources of air pollution, whereas, in rural areas, biomass burning for cooking and keeping warm generate air pollution. Apart from that, large-scale crop residue burning in agriculture fields, a cheaper alternative to mechanical tilling, is a major source of smoke, smog, and particulate pollution, especially in autumn and winter months. The Air Act (1981) has failed

DOI: 10.4324/9781003383970-20

to reduce pollution because of poor implementation of the rules. The Government of India, together with IIT Kanpur, in 2015 launched the National Air Quality Index. 'The National Clean Air Programme' with a tentative national target of 20–30% reduction in $PM_{2.5}$ and PM_{10} concentrations had been launched in India in 2019.

The Asansol subdivision of Paschim Bardhaman district of West Bengal has been declared by the Central Pollution Control Board (CPCB) of India as one of the 22 critically polluted areas of the country. This situation is an outcome of widespread industrialization, urbanization, and large-scale mining activities that have been taking place in the area for a long time without adequate environmental planning, according to the Asansol Durgapur Development Authority (ADDA, 2015). A considerable number of air polluting industries and collieries are located in Asansol subdivision. Primarily from industrial processes, airborne particulates pose a significant influence on atmospheric phenomena, plants, property, human and animal health. These activities produce dust particles and particulate matter that are suspended in the air. The machines used in cutting and removal of coal and waste materials are powered by diesel, the combustion of which emits carbon dioxide and sulphur dioxide in the air. Further, methane trapped in the subsurface coal seams gets released and mixes with the air. These suspended particles and poisonous gases are the main cause of air pollution in this geographical area.

The present study has been conducted over some time to quantify the spatial and seasonal variation in air pollutant concentrations and to measure the status of ambient air quality in the study area. Ambient air quality status was evaluated to monitor concerning particulate matter ($PM_{2.5}$ and PM_{10}), SO_2, and NO_2 over some time (2010–2019). Three different sites have been selected, viz., Asansol, Raniganj, and Jamuria for intensive analysis. Air quality monitoring during the pre-monsoon, monsoon, and post-monsoon seasons has been done to understand the seasonal variations.

The Study Area

The present study area, Asansol, is an administrative subdivision of the Paschim Bardhaman district in the state of West Bengal, India (Figure 16.1). It is located at 23°68′ N and 86°98′ E and having total of 831.89 km^2 (321.19 sq mi) area. It is the extended part of the Chota Nagpur Plateau in eastern India. It is undulating terrain with hard rock and laterite soil cover on it. The area is bounded by the river Ajoy on the north, river Damodar on the south, and river Barakar on the west. The area was heavily forested in the past. The discovery of coal in the eighteenth century led to the industrialization of the area and most of the forests have been cleared.

Asansol subdivision consists of 10 police stations, 4 community development blocks, 4 *panchayat samitis*, 35 *gram panchayats*, 181 mouzas, 165

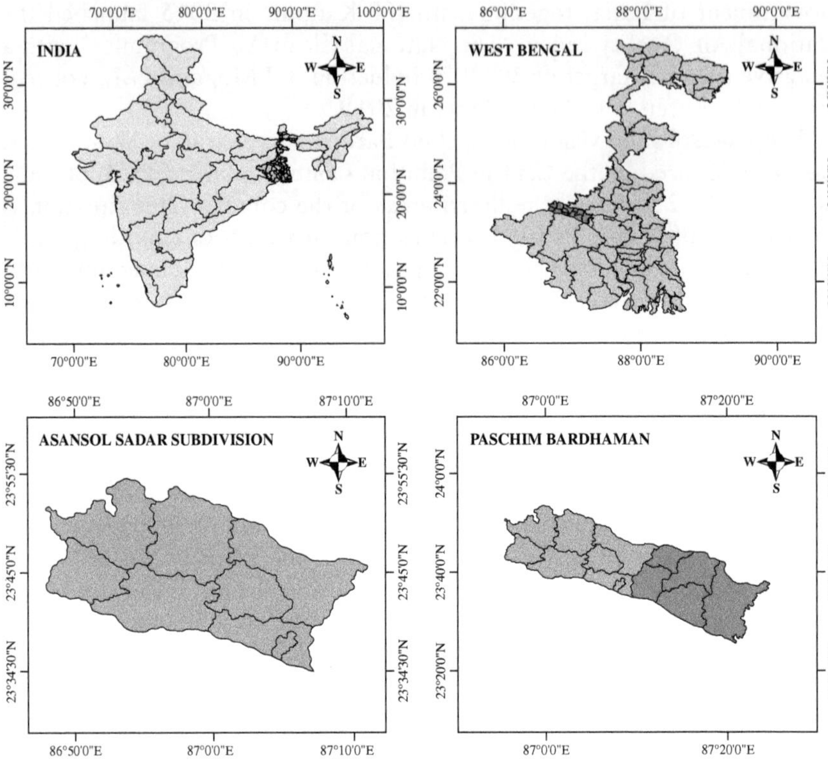

Figure 16.1 Location of the study area. Source: Prepared by the author.

inhabited villages, 1 municipal corporation, 3 municipalities, and 26 census towns. The only municipal corporation is the Asansol municipal corporation.

Important Sources of Air Pollution in the Study Area

Asansol is the second-largest city in West Bengal and many industries and coal mines are situated in this area. It is also a developing area and pollution is the necessary evil of all developmental activities. The principal sources of air pollution in this area are industries, vehicular exhaust, and mining activities.

Objectives of Study

The present study is an investigation to highlight the air quality status in Asansol subdivision. The main objectives of the present study are as follows:

- to analyze the trend of air quality in the Asansol subdivision.
- to enquire about the perception of local people regarding the air quality.
- to suggest some sustainable measures for maintaining good air quality.

Methodology

The present study is a combination of primary and secondary data analysis. An analytical methodology has been adopted to achieve the objectives. The progress of the work is made following the stages (Figure 16.2).

A set of primary data have been collected by surveys conducted in the study area by random sampling method. The survey has been carried out for a sample size of 247 at 4 CD Blocks of Raniganj, Jamuria, Barabani, and Salanpur and in Asansol Municipal Corporation in Asansol subdivision. The qualitative data collected by questionnaire surveys have been very much useful in understanding the perception of local people about the air quality in the study area.

The secondary data was collected from the West Bengal Pollution Control Board website. For a comparative study of air quality status in the Asansol subdivision area, the air quality-related data sets are downloaded in three different areas (Asansol, Raniganj, and Jamuria) in a period from 2006 to 2019 for Asansol, 2010 to 2019 for Raniganj and Jamuria. All collected data was then tabulated and used to develop different charts, analysis, and diagrams using statistical software. After careful observation of all outcomes, necessary concluding remarks have been made.

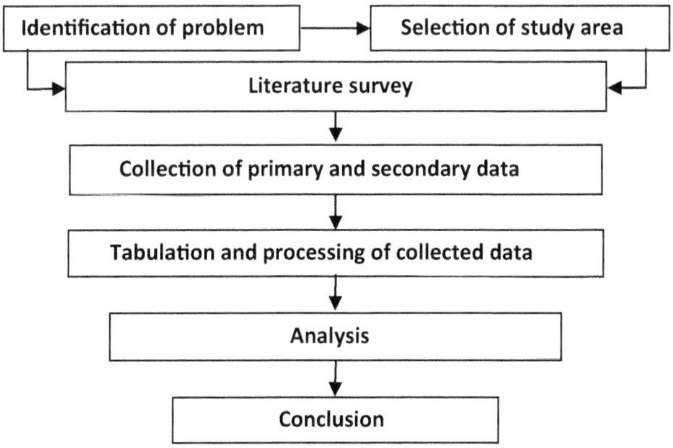

Figure 16.2 Flowchart of the methodology. Source: Prepared by the author.

Monthly Variation of Air Pollutants in Asansol Subdivision

A detailed analysis was carried out for the assessment of ambient air quality concerning suspended particulate matter (PM_{10}), sulphur dioxide (SO_2), and nitrogen dioxide (NO_2) at three sites Asansol, Raniganj, and Jamuria in Asansol subdivision from 2010 to 2019, with the help of data provided by the West Bengal Pollution Control Board.

Nitrogen dioxide (NO_2): The monthly average concentration of NO_2 was below National Ambient Air Quality Standards (NAAQS) (80 μg/m^3, CPCB 2009) guidelines of India from 2010 to 2019, except January and February in 2010 at Asansol, Raniganj, and Jamuria (Figure 16.3). During this period of time, the concentration of NO_2 was much higher in 2010 compared to 2019 but it also started slightly rising from 2017 to 2019. A trend of average monthly NO_2 concentration has been noticed that the amount was a little higher in winter seasons (November–February) in comparison to summer and rainy seasons.

Sulphur dioxide (SO_2): The average SO_2 concentration levels did not exceed the NAAQS (μg/m^3, CPCB 2009), at any of the sites. The average SO_2 levels were relatively higher in winter (November–February) than the summer (March–May) and the rainy season (June–September). The concentration of SO_2 has an increasing trend from 2010 to 2019. The lowest concentration was noticed in 2011 and the highest concentration was noticed during 2019, at all the sites (Figure 16.4). The all average SO_2 level was increased day by day at all the sites due to indiscriminate open-air burning of coal by the local inhabitants for cooking, vehicular traffic, etc. (Reddy & Ruj, 2003).

Particulate matter (PM_{10}): It was observed that the monthly average concentrations of PM_{10} at Asansol, Raniganj, and Jamuria from 2010 to 2019 were exceeding the limits of NAAQS (100 μg/m^3, CPCB 2009). The average monthly concentration was found to be much higher during winter (November–February) in comparison with the summer (March–May) and monsoon (June–September). From June to September in 2010–2015, the average concentration level did not exceed the NAAQS in the three sites (Figure 16.5).

The trend of PM_{10} concentration is increasing day by day due to accelerated industrial, mining, and other community activities and also due to increased vehicular traffic (Reddy & Ruj, 2003).

Seasonal Variation of Air Quality in Asansol Subdivision

There is a significant variation in air pollution levels with the changing season. The concentration of particulate matter increases during winter followed by summer and rainy seasons in the study area. Similar trends

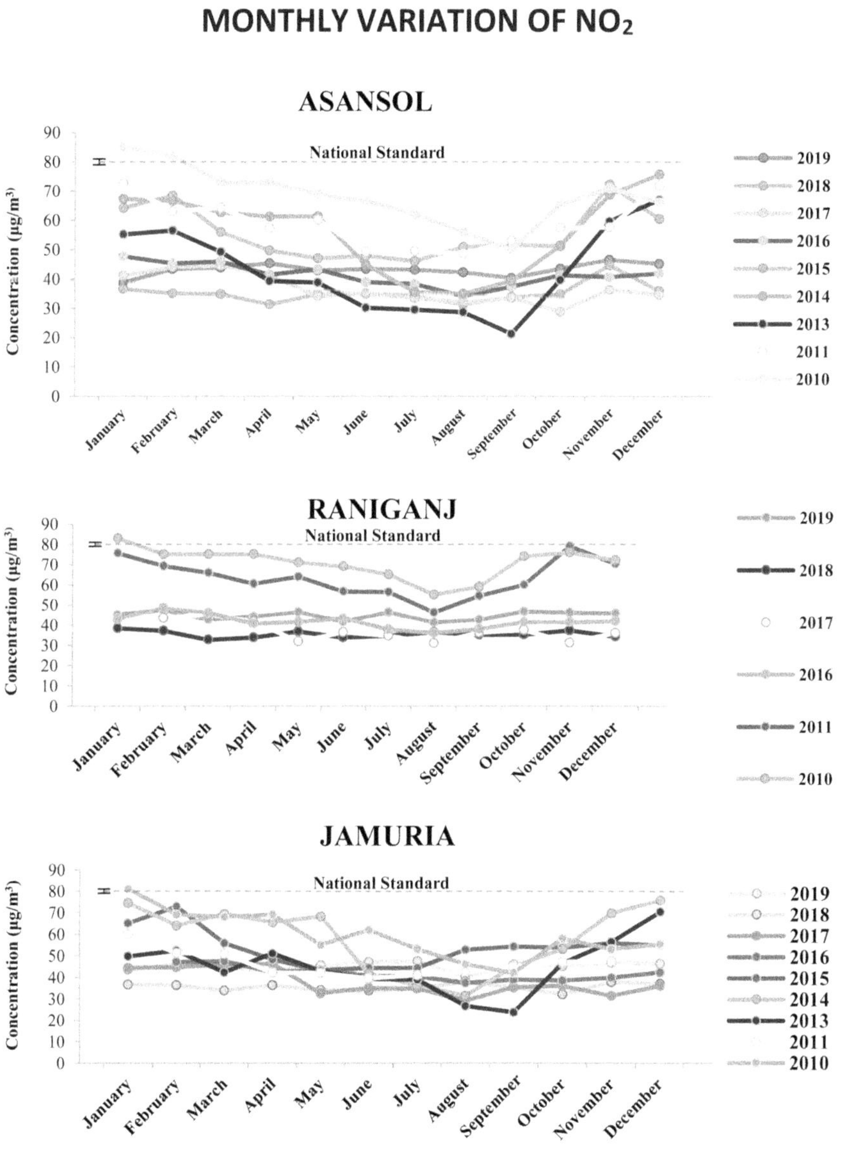

Figure 16.3 Monthly variation of NO_2. Source: Data from West Bengal Pollution Control Board.

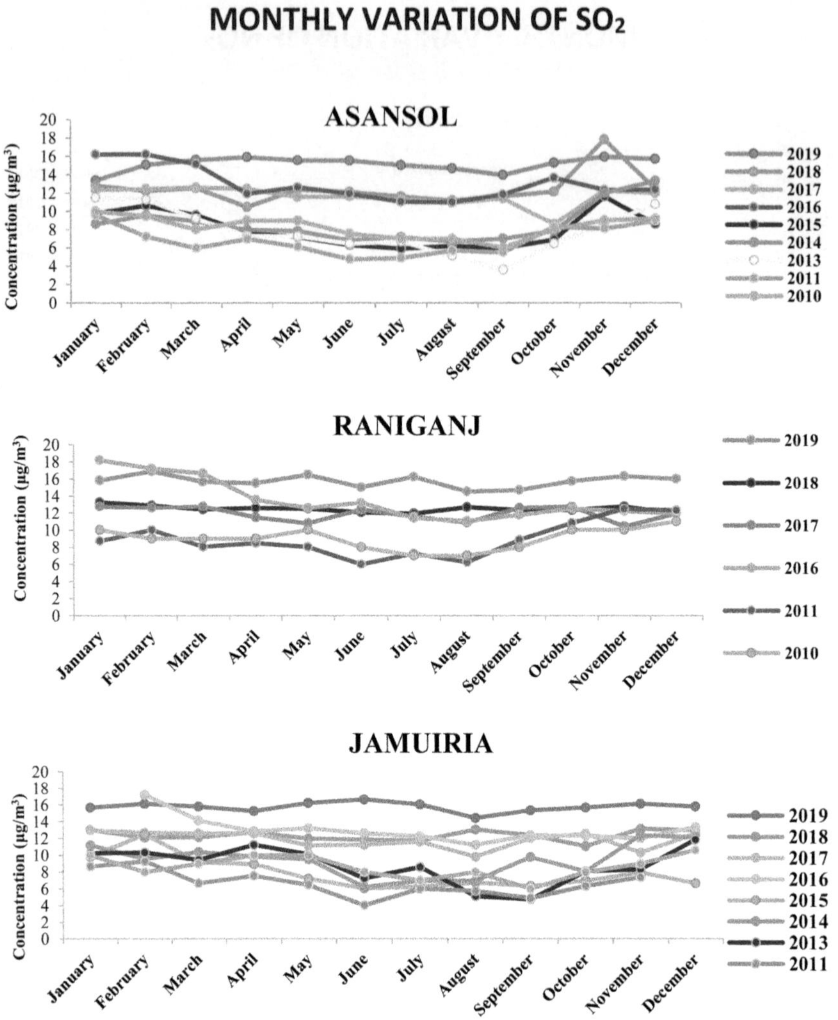

Figure 16.4 Monthly variation of SO_2. Source: Data from West Bengal Pollution Control Board

are identified in different areas of India such as Jharia coalfield in Odisha (Pandey, Agrawal, & Singh, 2014), Raniganj–Asansol area (Reddy & Ruj, 2003). The primary source of fugitive dust at fully operational surface mines may include overburden removal, mineral stockpiles, and site restoration (Appleton, Kingman, Lowndes, & Silvester, 2006).

NO_2 concentration: The concentration of NO_2 was below the NAAQS (80 µg/m³) guidelines of India during all the seasons and at all the sites under

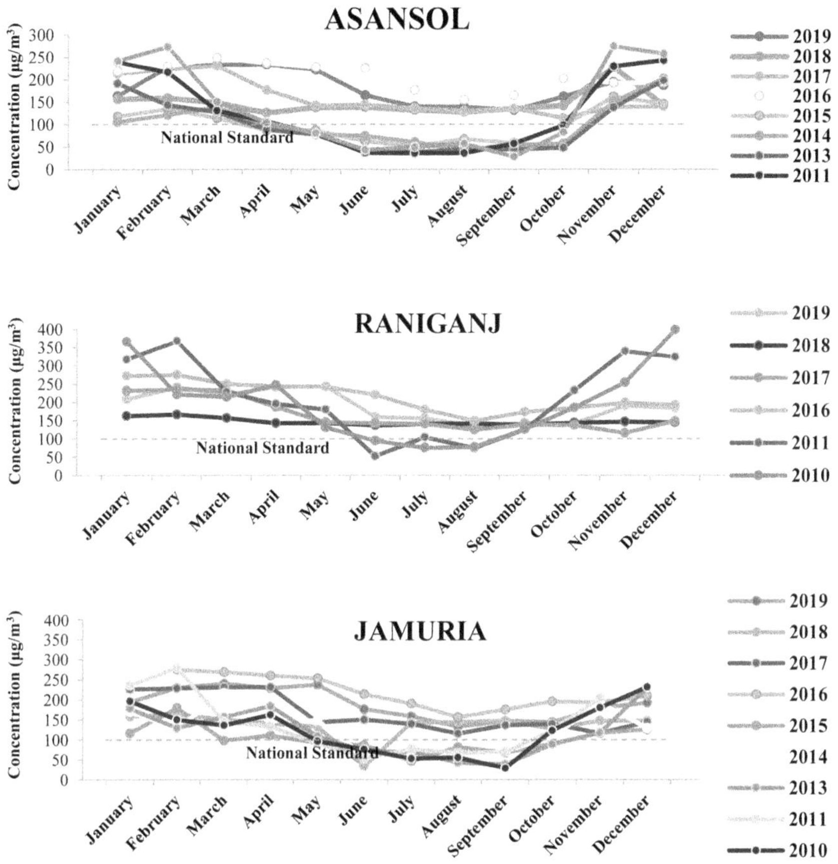

Figure 16.5 Monthly variation of PM_{10}. Source: Data from West Bengal Pollution Control Board.

study (Figure 16.6). The nature of mean concentration of NO_2 during the summer season was maximum in Jamuria and least in Asansol, during the rainy season it was maximum in Raniganj and Asansol was the least polluted. During winter the Asansol became the highest polluted place followed by Raniganj and Jamuria in 2018. The concentration of NO_2 had increased in the year 2019 (Figure 16.6). During the summer and winter seasons, concentration of NO_2 was maximum in Raniganj followed by Jamuria and Asansol. In the rainy season too, Raniganj remained the most polluted place, followed by Asansol and Jamuria. Various anthropogenic and industrial emissions like burning activities, including vehicle exhaust, coal, oil, and natural gas, blasting in coal mines, etc. are the main source of NO_2 concentration in air in

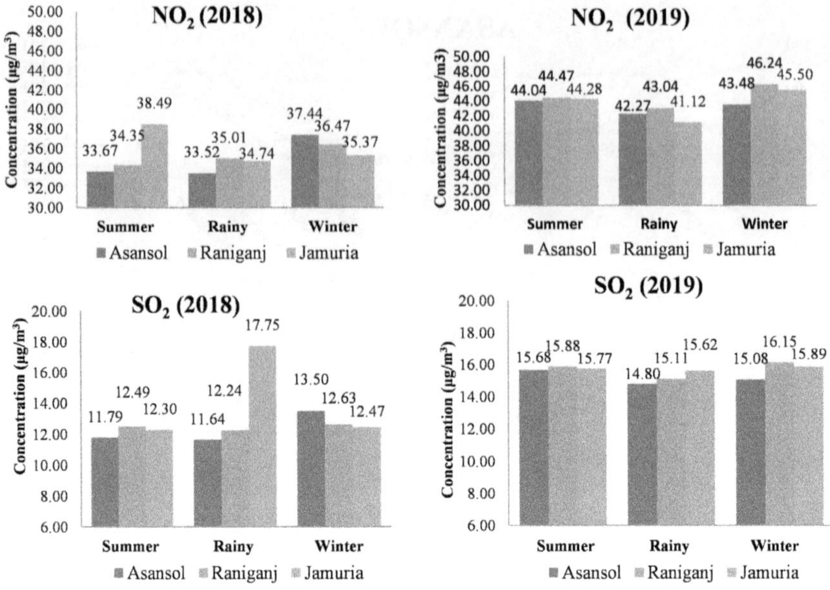

Figure 16.6 Seasonal variation of SO_2 and NO_2 concentration. Source: Data from West Bengal Pollution Control Board.

the present study area. The seasonal variation in NO_2 concentrations is mainly determined by the varying day length over the year. In absence of sunlight, NO_2 has a longer lifetime in the atmosphere (Li, Matthews, & Sinha, 2007), which explains the reason for higher NO_2 during winter, which leads to increased levels of NO_2 during this season (Atkins and Lee, 1995).

SO_2 concentration: The concentrations of SO_2 were below NAAQS (80 µg/m³) guidelines of India during all the seasons and at all sites (Figure 16.6). SO_2 concentration during the summer season was maximum in Raniganj followed by Jamuria and Asansol, during the rainy season Jamuria was the most affected place followed by Raniganj and Asansol. In the 2018 winter, Asansol was the most polluted place followed by Raniganj and Jamuria. The SO_2 concentration increased during 2019. It has been noticed that SO_2 concentration in the air of the study area remains maximum in winter and gradually decreases from summer to rainy season. Wind speed, air masses, and height of the mixing layer also govern SO_2 concentration in air.

PM_{10} concentration: The concentrations of PM_{10} at Asansol, Raniganj, and Jamuria during winter of 2018 were, respectively, 186.25, 192.77, and 188.01 µg/m³ and were 1.86, 1.92, and 1.88 times higher than

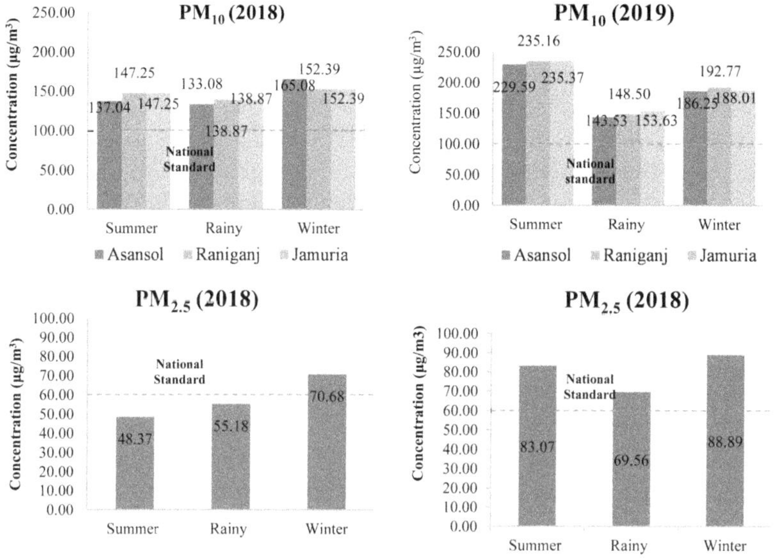

Figure 16.7 Seasonal variation of PM_{10} and $PM_{2.5}$ concentration. Source: Data from West Bengal Pollution Control Board.

NAAQS (100 µg/m³, CPCB 2009) guidelines of India (Figure 16.7). During summer 2018, PM_{10} concentrations were 1.37, 1.47, and 1.47 times higher than NAAQS at respective sites. During the rainy season of 2018, the concentrations of PM_{10} were 1.33, 1.38, and 1.38 times higher than NAAQS at respective sites. The trends of PM_{10} concentration was a slight increase during 2019 (Figure 16.7). The concentrations of PM_{10} during summer were 229.59, 235.16, and 235.37 µg/m³ and were 2.29, 2.35, and 2.35 times higher than the NAAQS in Asansol, Raniganj, and Jamuria, respectively. During the rainy season of 2019, PM_{10} concentration slightly decreased; during the winter season, it was increased again.

Throughout these seasons the PM_{10} concentrations were 1.86, 1.92, and 1.88 times higher than NAAQS's level in Asansol, Raniganj, and Jamuria, respectively. PM_{10} concentration varied significantly in different seasons and it has an increasing trend. The coarse particles (PM_{10}) are formed due to mechanical disruption (e.g. crushing, grinding, and abrasion of surfaces), evaporation of sprays, and re-suspension of dust (Sastry, Chansar, Nagesha, Muralidhar, & Mohiuddin, 2015) near coal mining areas; the coal transportation was the main source of such pollution (Chaulya, 2005) and hence proportions of coarse particles were high in the study area.

$PM_{2.5}$ concentration: The concentration of $PM_{2.5}$ in Asansol during winter of 2018 was 70.68 μg/m^3; it was 1.17 times higher than the NAAQS (60 μg/m^3, CPCB 2009) limit in India. During the summer and rainy season of 2018, $PM_{2.5}$ concentrations were, respectively, 48.37 and 55.18 μg/m^3 at that place. However, the concentration of $PM_{2.5}$ increased during 2019. During summer, the concentration was 1.38 times higher compared to the NAAQS, during the rainy season it was 1.15 times higher than the NAAQS's limit, and during winter it was 1.48 times higher than the NAAQS at the site (Figure 16.7). $PM_{2.5}$ concentrations showed significant variation and increasing nature over seasons. Such particles are formed by chemical reaction, nucleation, condensation, coagulation, evaporation of fog and cloud droplets, in which gases also dissolve and react (Wilson & Suh, 1997). The burning of coal is a major source of fine particles. Active mine fire and the industrial emission present in the study area play a major role in the generation of $PM_{2.5}$. There is a very small difference in the concentration of $PM_{2.5}$ during the summer and winter seasons (Figure 16.7). This may be attributed to the fact that $PM_{2.5}$ remains airborne through nonlinear processes for days to weeks during monsoon months as washout processes are least effective for cleansing these particles.

A higher concentration of particulate matter during winter can be attributed to low temperature and low wind speed, which lead to lower mixing height and poor dispersion conditions. During the summer season, particulate matter concentrations were found to be lower than winter due to enhanced dispersion caused by high wind speed. The lowest concentration was observed during the monsoon season, because of washout by rainfall and also due to higher relative humidity, which reduces re-suspension of dust (Pandey, Agrawal, & Singh, 2014).

Annual Trend of Air Pollutant Concentration

The time series analysis has been conducted for assessment of the trend of annual average air pollutant concentration at Asansol, Raniganj, and Jamuria in the Asansol subdivision area from 2010 to 2019. Following results came out from the analysis.

The trend of NO_2 concentration in Asansol, Raniganj, and Jamuria from 2010 to 2019 states that its concentration was slightly decreased at all the stations. The trend of SO_2 concentration was increasing significantly in all sites. This increasing rate is much higher in Raniganj areas as compared to Asansol and Jamuria. The PM_{10} concentration rate is increasing in Asansol and the Jamuria region. Its concentration at those sites has an upright trend from 2010 to 2019, but in Raniganj the trend was slightly decreasing. The trend model of SO_2 shows an increased forecast for the next five years in the study area (Figure 16.8). The PM_{10} concentration trend is slightly decreasing in the same time series model.

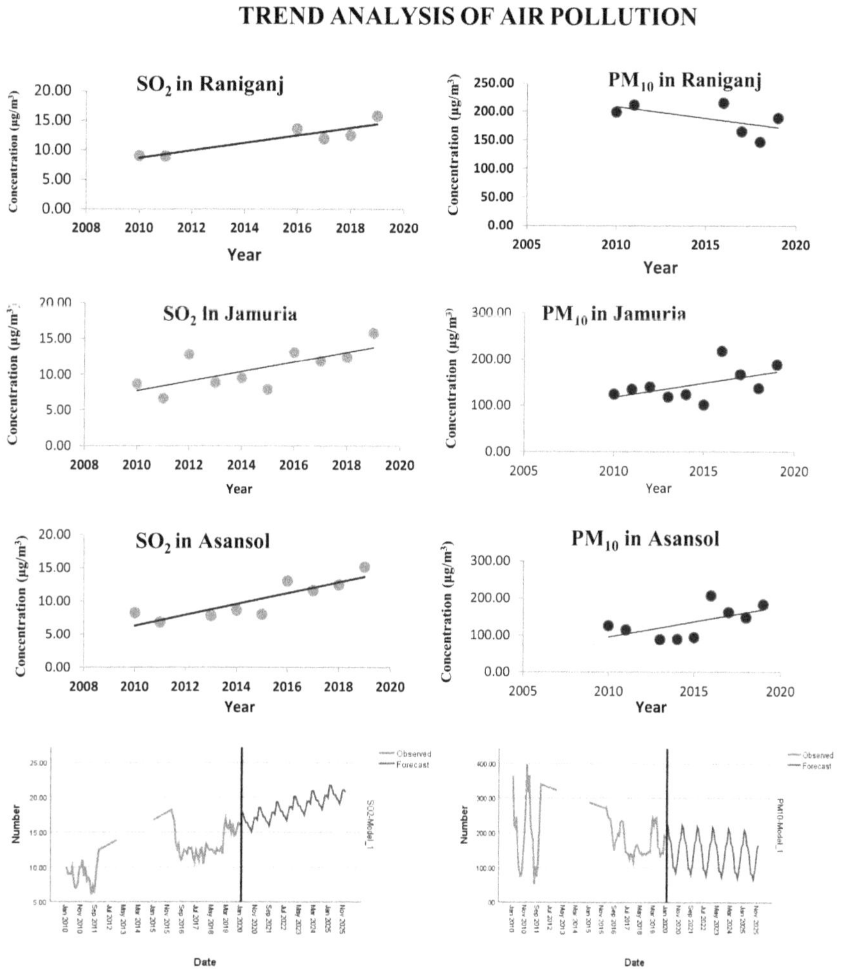

Figure 16.8 Trend analysis of air pollution. Source: Data from West Bengal Pollution Control Board.

Environmental Perception of Respondents about Air Quality

From the above discussion, it is quite evident that the air quality of the Asansol subdivision is highly alarming. Poor air quality may trigger many environmental problems and health hazards to the residents. But an environmental study remains incomplete until the perception of local people regarding the problem is not addressed. With this aim, a perception study has been done. Its salient findings are discussed here. This is also helpful in developing a strategy for sustainable environmental management.

Responses have been collected from five blocks of Asansol subdivision, viz., Asansol, Raniganj, Jamuria, Barabani, and Salanpur. It is understood that environmental problems are very common in this region (Figure 16.9). Almost in all the blocks respondents strongly agree that environmental problems in Asansol subdivision are very common.

One hundred per cent of respondents in Salanpur strongly agree that people are suffering from air pollution due to collieries and heavy industries. On that issue, 80%, 65.22%, 62.86, and 45.46% of respondents strongly agreed in Barabani, Raniganj, Asansol, and Jamuria, respectively (Figure 16.10).

Most of the respondents are not satisfied with the air quality. In Asansol, 34.29% of respondents strongly supported that. In Raniganj, 32.61% of respondents strongly agreed with that. The majority of respondents in Jamuria, Barabani, and Salanpur region supported the notion too (Figure 16.11). As the secondary data shows, air pollution is on the rise in the region and public

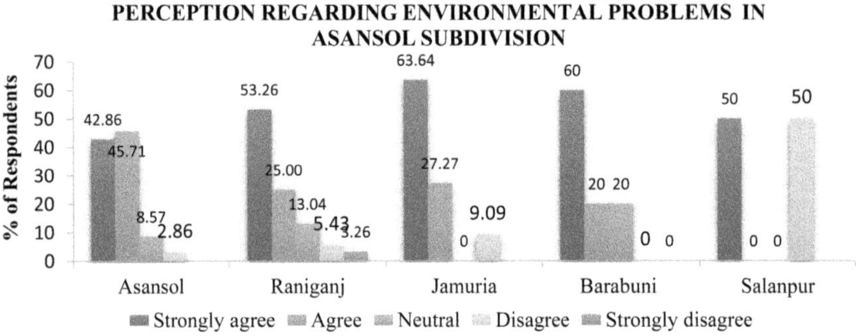

Figure 16.9 Perception regarding environmental problems in Asansol subdivision. Source: Primary data.

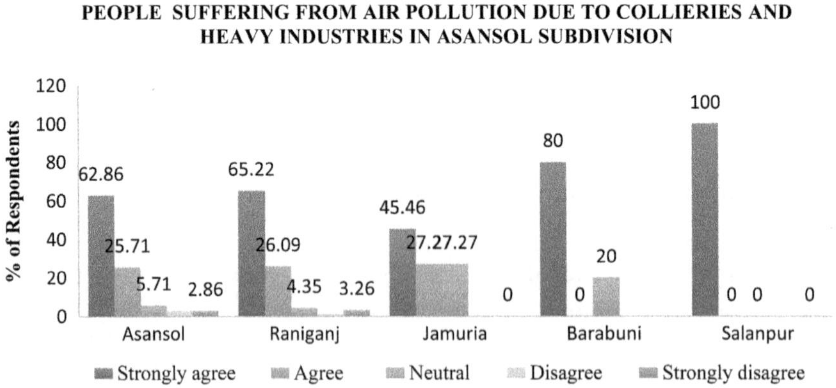

Figure 16.10 People suffering from air pollution due to collieries and heavy industries in Asansol subdivision. Source: Primary data.

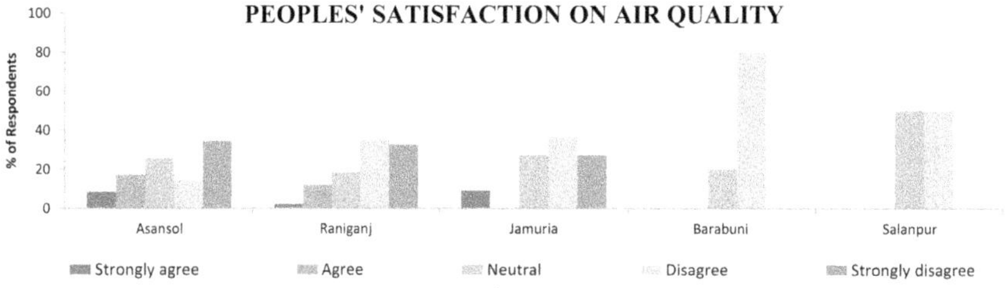

Figure 16.11 Peoples' satisfaction on air quality. Source: Primary data.

perception also supports that fact. Though few of the respondents remained neutral in that respect, the general trend of people confirmed poor ambient air quality. There is an overall grudge of people against the administration for failing to control the deterioration of air quality. People strongly believe that the administration should take effective measures to improve the condition. Some of the respondents feel that the local administration is working on it, though it is not enough. More efforts still are to be taken.

Air pollution is considered as one of the serious environmental risk factors in the study area and it causes some diseases such as asthma, lung cancer, ventricular hypertrophy, and Alzheimer's. Air pollution is also associated with increased anxiety and mental disorders, such as depression, schizophrenia, and autism. From the opinions of local people in the Asansol subdivision, it has been noticed that the air pollution has already started affecting the social life and they believe that it will also affect the future generation in that region. About 28.57% respondents of Asansol, 21.74% respondents of Raniganj, 45.45% respondents of Jamuria, 40% respondents of Barabani, and 20% respondents of Salanpur strongly agree with that. Forty per cent respondents of Asansol, 43.48% respondents of Raniganj, 27.27% respondents of Jamuria, 40% respondents of Barabani, and 80% respondents of Salanpur agree that the air pollution is already affecting their society.

However, about 48.57% respondents of Asansol, 38.04% respondents of Raniganj, 72.73% respondents of Jamuria, 60% respondents of Barabani, and 80% respondents of Salanpur block strangely agree that the air pollution will impact the future generations of the Asansol subdivision region (Figure 16.12). So the local people think that the effects of air pollution will increase in the future and it is harmful to future generations of the Asansol subdivision area. This mass opinion also confirms the trend observed in secondary data analysis.

Air pollution not only contributes to respiratory diseases in humans, but it can also affect plants and animals of the region. The effects of air pollution on plants develop over time. The impact of air pollution on animals is more or less similar to that on man. To control the situation, both the

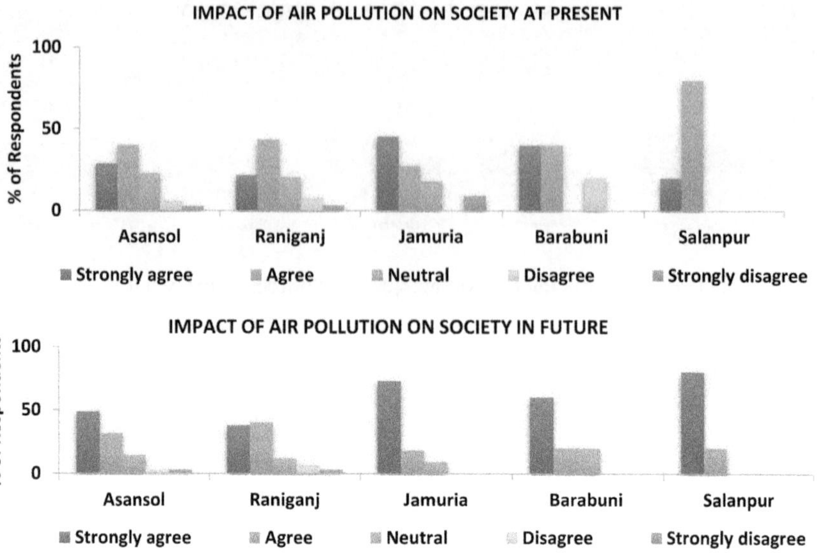

Figure 16.12 Impact of air pollution on society in present and future. Source: Primary data.

government and civilians have an important role to play, but the government has a much greater responsibility than an individual regarding the formulation and implementation of a proper plan or law.

Sustainable Measures to Maintain the Air Quality

The air quality status in the Asansol subdivision is worse than the national standard. The concentration of pollutants in the ambient air is continuously increasing due to industrial outputs, vehicular exhausts, and mining activities. If an appropriate procedure is not being taken now, then it would become harmful for the natural environment as well as for human health. Considering this, the following measures might be helpful to improve the condition:

- more use of non-conventional energy sources compared to fossil fuels.
- increased use of public transport.
- encouraging afforestation over the wastelands associated with collieries.
- burning of agricultural and other wastes should be controlled.
- restriction over unauthorized open cast mining.
- regular monitoring of air quality.
- installation of some pollution control devices, including gravitational settling chamber, centrifugal collectors, wet collectors, electrostatic precipitators, and fabric filters, in the industries.

Conclusion

Assessment of air quality status in the Asansol subdivision area has shown that the present condition of air quality is not satisfactory in that area. The data provided by West Bengal Pollution Control Board helped to analyze the trend and present condition of air quality in the study area. Though its ill effects are still not so prominent, it is harmful to future generations and the natural environment. The result of the study indicates that the concentrations of all major pollutants have been increasing during the study period, except NO_2 concentration. The concentration of NO_2 at Asansol, Raniganj, and Jamuria areas is showing a slightly decreasing trend. SO_2 concentration in that region was not too high but it has an increasing trend from 2010 to 2019. The PM_{10} and $PM_{2.5}$ concentrations are much higher and it has also increased significantly from 2010 to 2019. These pollutants mostly originate from the emissions of various mining and industrial activities. The annual and daily average concentrations of PM_{10} and $PM_{2.5}$ were higher than the NAAQS at Asansol, Raniganj, and Jamuria areas. This is one of the reasons behind the decline in the air quality in the region. The seasonal variation of air quality is also a remarkable feature in that region. It has been observed that the concentrations of pollutants are higher in winter in comparison to the summer and the rainy seasons. Wind speed, relative humidity, temperature, and rainfall are the important controlling parameters of seasonal variations in air pollutant concentration in this area.

From the perception study regarding the ambient air quality of the study region, it is clear that the air quality in the region is not satisfactory for the local inhabitants, even for the entire living components of the environment. The maximum numbers of respondents agree that environmental problems are very common in the area. The inhabitants are suffering due to unhealthy air quality near coal mines and heavy industrial units. Even they agree that the air pollution in this region is slowly but steadily increasing which is not at all desirable for future generations and it is also harmful to local flora and fauna.

It may be concluded that due to the huge quantity of PM_{10} and $PM_{2.5}$ concentrations in the ambient air, the quality of air in the Asansol subdivision area has crossed the threshold of the standard limit. It is very challenging to maintain good air quality at the regional level because this area has a long history of mining and industrial activities. But some control measures at the administrative and local level are needed at the site to reduce the generation of particulate matter at the source. A green belt may be developed over the abandoned collieries and around heavy industries. With the strict implementation of additional control measures at the sensitive sites, the air quality in the study area could be brought within the national ambient air quality threshold limit. For this, strong cooperation between the administration and local people is required.

Works Cited

ADDA. (2015). *Land Use and Development Control Plan - 2025 for Asansol Sub-Division*. Asansol: Asansol-Durgapur Development Authority

Appleton, T., Kingman, S., Lowndes, I. and Silvester, S. "The Development of a Modelling Strategy for the Simulation of Fugitive Dust Emissions from in-pit Quarrying Activities: A UK Case Study." *International Journal of Mining Reclamation and Environment*. 20 (2006): 57–82.

Bergstra, A.D., Brunekreef, B. and Burdorf, A. "The Effect of Industry-related Air Pollution on Lung Function and Respiratory Symptoms in Schoolchildren." *Environmental Health*. 17 (2018): 30. http://doi.org/10.1186/s/12940-08-0373-2 Accessed 4th Dec,2019 9.45pm

Chaulya, S. "Air Quality Status of an Open Pit Mining Area in India." *Environmental Monitoring and Assessment*. 105 (2005, July): 369–389.

EEA. *Air Quality in Europe*. Luxembourg: Publications Office of the European Union, 2013.

Goyal, P. and Sidharth. "Present Scenario of Air Quality in Delhi: A Case Study of CNG Implementation." *Atmospheric Environment*. 37.38 (2003, December): 5423–5431.

Guha, D. "A Case Study on the Effects of Coal Mining in the Environment Particularly Concerning Soil, Water, and Air Causing a Socio-economic Hazard in Asansol-Raniganj Area, India." *International Research Journal of Social Sciences*, 3.8 (2014, August): 39–42.

Jiming, H. and Guowen, L. "Point Sources Of Pollution: Local Effects and It's Control – Vol. I - Air Pollution Caused by Industries." *Encyclopedia of Life Support Systems (EOLSS)*, 171–178.

Li, S., Matthews, J. and Sinha, A. "Atmospheric Hydroxyl Radical Production from Electronically Excited NO2 and H2O." *Science*.318(5856) (2007): 1657–1660.

Mage, D., Ozolins, G., Peterson, P., Webster, A., Orthofer, R., Vandeweerd, V., et al. "Urban Air Pollution in Megacities of the World." *Atmospheric Environment*. 30 (1996): 681–686.

Nagdeve, D. "Environmental Pollution and Control: A Case Study of Delhi Mega City." *Population and Environment*. (2004, May): 25–29.

Pandey, B., Agrawal, M. and Singh, S. "Assessment of Air Pollution Around Coal Mining Area: Emphasizing Spatial Distributions, Seasonal Variations, and Heavy Metals, Using Cluster and Principal Component Analysis." *Atmospheric Pollution Research*. 5(1) (January 2014): 79–86.

Pio, C. and Feliciano, M. "Dry Deposition of Ozone and Sulphur Dioxide over Low Vegetation in Moderate Southern European Weather Conditions: Measurements and Modelling." *Physics and Chemistry of the Earth*. 21 (1996): 373–377.

Rathi, A. "Promotion of Cleaner Production for Industrial Pollution Abatement in Gujarat (India)." *Cleaner Production*. 11(5) (Aug 2003). 583–590.

Reddy, G. and Ruj, B. "Ambient Air Quality Status in Raniganj-Asansol Area, India." *Environmental Monitoring and Assessment*. 89 (2003): 153–163.

Saini, V., Gupta, P.R. and Arora, M.K. "Environmental Impact Studies in Coalfields in India: A Case Study from Jharia Coal-field." *Renewable and Sustainable Energy Reviews*. 53 (January 2016): 1222–1231.

Sastry, V., Chansar, K.R., Nagesha, K., Muralidhar, E. and Mohiuddin, M.S. "Prediction and Analysis of dust Dispersion for Drilling Operation in Opencast coal Mines." *Procedia Earth and Planetary Science.* (2015): 303–311.

Schwegler, F. "Air Quality Management: A Mining Perspective." *WIT Transactions on Ecology and the Environment.* 86 (2006): 205–212.

Trivedi, R., Chakrabarty, M. and Tewary, B. K. "Dust Dispersion Modeling Using Fugitive Dust Model at an Opencast Coal Project of Western Coalfields Limited, India." *Journal of Scientific & Industrial Research.* 68 (2009, January): 71–78.

West Bengal Pollution Control Board. (2000). *West Bengal Pollution Control Board.* Retrieved from wbpcb: http://www.wbpcb.gov.in/

Wilson, W.E. and Suh, H.H. "Fine Particles and Coarse Particles: Concentration Relationship Relevant Epidemiologic Studies." *Journal of the Air & Waste Management Association.* 47(12) (1997): 1238–1249.

17 Tracing Nature's Footprints in English Literature

An Ecocritical Perspective

Arunima Karmakar

I

Ecocriticism, as defined by Cheryll Glotfelty in the introduction to *Ecocriticism Reader*, is the study of the inter-relationship between physical environment and literature from a multidisciplinary point of view. Since time immemorial, Nature has been omnipresent in literature, as it had been the most fascinating subject matter for authors. Literary texts, like the naturalists and conservationists, analyzed environmental issues, treating nature as a subject, to illuminate the readers about the various ecological threats, their possible solutions, and thereby trying to redress the balance of Nature on Earth, which has slipped too far towards the direction of extinction. Since its inception in the mid-1990s, ecocritics have been engaged in issues of contemporary political and social awareness surrounding anthropocentric ecological degradation and climate change. Like a veritable monarch, Nature rules over man and has an indelible impact on human life. It is the source of happiness and accomplishment, as it motivates man and helps combat depression and sorrows by developing a sense of aesthetic beauty. Nature plays a didactic role by inspiring us to accept the vagaries of human existence. Ecocriticism is a postmodern study as ecological concerns have taken centre stage in social and literary spheres. It focuses on the destabilization of nature owing to the accelerating urbanization and industrial capitalism. With the change in the human world there has been a change in the representation of nature in literary works down the memory lane. This chapter seeks to embark upon a retrospective literary journey that bears witness to an unending ecocritical awareness through an analysis of representation of nature in literary works and find out if the truth of ecology lies in literature or not. This chapter analyzes nature as represented in some seminal texts down the centuries (prior to the inception of ecocriticism) to the present times to explore the change in the representation due to the changes in human history.

From the inception of language to contemporary times, Nature is discernible in literature as also in human life. Its awe-inspiring omnipotence was interpreted mythically to represent its all-pervading persona. Nature was

DOI: 10.4324/9781003383970-21

a spectacle to be admired as well as a medium of communication between the gods and man. All unusual and violent phenomena of nature seemed to convey divine warning, wrath, or prophecy. Ancient Greek epics like Homer's *The Iliad* and *The Odyssey* and Virgil's *Aeneid* are replete with beautiful, vivid descriptions of the natural world. These descriptions, ranging from the Asian meadows on the banks of Kaystrios, the wine-dark sea, the rosy-fingered dawn to the grooves of the nymphs of Ithaca, haunt of geese, cranes, and swans, Sicilian shores, the Tiber, stir the heart of the modern nature lover and ecocritics. Even in the Bible, nature plays a vital role in man's understanding of God's nature. It is through nature that God ordains us to protect all his creations, which forms the crux of Christian faith. Indian epics, *The Ramayana* and *The Mahabharata*, uphold the ancient Indian culture of life in unison with cosmic rhythm through the unique concept of 'Dharma.' These texts reverberate with the celebrations of nature as an inseparable part of human existence. Our Vedic texts, the Gurukul and Ashrama, the first preparatory schools infuse in the learners an awareness of worshipping, nourishing, and cherishing nature in our everyday life. The world then was a veritable paradise with peace and harmony. Although the genesis of these ancient texts did take place in a pre-industrialized world, we find that there were strictures against polluting water resources and instructions of afforestation. We can trace the dominant role played by nature in human life as represented in these ancient classic texts before scientific intervention.

II

As the wheels of civilization rolled on, human life seemed to get complicated and this found reflection in literature too. Life became contaminated and complex. Man's insatiable desire started to exploit nature mercilessly like a ruthless monarch. The Old English epic *Beowulf* presents nature as hostile, reflecting demonic antagonism. It signifies how human artifice, like armour and adornments, helps man to exert control over the chaotic natural world. Beowulf's fight with the monster symbolizes the human desire for victory over the uncontrollable and malevolent natural forces. The exterior landscape represents the mentally structured space to be the clue to understand the psychosis of the characters. Another Old English text, *The Seafarer*, is an exultation of boundless nature with its presentation of blossoming trees and blooming fields urging the Seafarer to wander afar over the tide-ways; while the cuckoo, the harbinger of summer, warns him with her moaning song which foretells sorrow.

In the Middle English period, the most notable work which is replete with references to nature is Geoffrey Chaucer's *The Prologue* to *The Canterbury Tales*. He begins with a eulogy to nature which was an accepted medieval convention:

> Whan that April with his shoures soote
> The droghte of March hath perced to the roote,

And bathed every veyne in swich licour
Of which vertu engendred is the flour.
(73)

Chaucer uses nature judiciously as a symbol for spiritual regeneration and biological procreation through the process of *natura naturata* and *natura naturans*.[1] The idyllic nature has been portrayed vividly in this work with references to the socio-political condition of fourteenth-century life with references to corruption in the ecclesiastical order, epidemics of Black Death[2] and the plague as also the first Peasant Revolt.[3] Nature imagery and animal imagery are used in profusion to add individual unique traits to most of the pilgrims while introducing them to the readers. For example, the character of the Prioresse is hailed by the name 'madame Eglentyne' which means wild rose, signifying the beauty of the lady and also obliquely pointing towards her forbidden passionate life. These texts reflect the disruption of nature by the heedless agriculture conducted by man which started to damage the natural world.

Nature seems to be one of the dominating spirits in the domain of Renaissance literature. The familiar voices of Spenser, Shakespeare, Jonson, and Marvell mingle with a diverse chorus of farmers, herbalists, shepherds, hunters, foresters, philosophers, sailors, sky-watchers, and duchesses, as well as ventriloquized beasts, trees, and rivers. Sonnet sequences of Sidney and Spenser testify to nature as a potent tool to express their love and fear of death. The 'Doctrine of Nature'[4] forms the core of Shakespeare's view of life. Like his contemporaries, he draws a lot from the natural world. He uses nature as a theatrical space in which some of the main actions of Shakespearean plays are set. Nature is used as a metaphor, a figurative landscape that expresses psychological, social, and emotional anxieties. In Shakespearian comedies, the most dominating and fascinating space is none other than the forest or the woodland, as seen in *As You Like It*, *A Midsummer Night's Dream*, and *The Two Gentlemen of Verona*. Nature here acts as a space for transformation and of magical aspiration, the fulfilment of fantasy, and festive escapism. In tragedies, like *King Lear*, *Macbeth*, *Julius Caesar*, and *Titus Andronicus*, nature is the place of destruction, brutality, and dangerous transformation. Nature plays a significant role in his plays as it amplifies the performance and figuratively adds to the realization of emotional and psychological destruction as is evidenced in *King Lear*. In Act I, Scene III of his Roman tragedy *Julius Caesar* in the speech of Casca, the description of natural elements in the apocalyptic night foretells the threats that were nursed by the conspirators against Julius Caesar:

Are you not moved, when all the sway of earth
Shakes like a thing unfirm? O Cicero,
I have seen tempests, when the scolding winds
Have rived the knotty oaks, and I have seen

The ambitious ocean swell and rage and foam
To be exalted with the threatening clouds.

(184–185)

Natural reverberations of human misappropriation form an important trope in Shakespeare's dramatic arena. The unnatural activities reflected in the natural world, on the night before Duncan's murder by Macbeth, are a case in point.

Seventeenth century saw a shift in the predominant British view of the natural world, from the position that God created all for human use to a greater emphasis on human stewardship and responsibility of it. John Milton in *Paradise Lost* through the fatalistic fall of Adam and Eve from Eden portrays the general plight of human beings under the demonic mechanistic materialism. He presents the contemporary scenario where man is suffering from a Satanic alienation from the natural world due to the concomitant rise of science. Adam and Eve's eating of the forbidden fruits testifies the unintended cannibalism that resides at the core of all humans and surfaces out as one tries to earn supremacy and power. Milton is thus considered to be the precursor of modern environmentalism as he interpreted the 'Fall of Man' being anchored by the early scientific thought of his time, as it separated human and the natural world. Thus, he advocated for proper human attitude towards the natural world. Men should play the role of a benevolent custodian of nature that God has conferred to man rather than be a ruthless exploiter or colonizer.

During the Restoration Age and the eighteenth century, Nature was methodized; premium importance was placed on human reason and on an empirical philosophy that held that knowledge about the world was through the senses and by applying reason to what we take in through our senses. The reason was an unchanging, uniquely human characteristic that served as a guide for man. The pseudo-classical writers of this period looked at Nature through the lenses of Juvenal, Horace, and Ovid. To them, to find delight in Nature was a sign of wanting in taste and culture. Nature to Alexander Pope meant reason and suggested that nature can impart life-force and beauty to art. In his *Pastorals*, Pope initiates the theory of a relationship between nature and art. He proclaims that:

Those Rules of old discover'd, not devis'd,
Are Nature still, but Nature methodiz'd;
Nature, like Liberty, is but restrained
By the same Laws which first herself ordained.

(120)

A distrust for sophistication and a love for simplicity found expression in the works of Romantic poets like Wordsworth, Shelley, Keats, Byron, and Blake who called for a return to nature for nature's sake. They were disillusioned with the French Revolution[5] and sought Nature for its 'healing power.' Wordsworth, the high priest of Nature, believed God and Nature to be one and comprehended Nature as a universal spirit and guiding force. These lines from "Lines Written a Few Miles above Tintern Abbey" depict his belief in the balance between landscape and mindscape:

> How oft, in spirit, have I turned to thee o Sylvan Wye!
> Thou wanderer through the woods,
> How often my Spirit turn to thee!
>
> (115)

Coleridge was enchanted with the mystic aspects of Nature as reflected in "Kubla Khan" and "The Rime of the Ancient Mariner." Shelley depicts the mystic bond between Nature and man in his mythopoeic vision. "Ode to the West Wind" depicts Nature as a destroyer and a preserver. The "Odes" of Keats are rich in sensuous appeal. Two poems by Blake with the same title named as "The Chimney Sweeper" present the deplorable socio-economic condition of workers in the eighteenth century endangered by the adverse effects of industrialization.

Victorian era reacted against Romanticism in which Nature was replaced by a Darwinian predator. This age witnessed the inexorable process of disintegration with dismay. While Wordsworth's construction of Nature is cultural and aesthetic, the Victorian construct of Nature is ecological. Victorian era places humanity within the frame of 'industrial capitalism,' where Anthropocene emanates as a new geologic age defined by human action, which is most often dated to the Industrial Revolution. Re-imagining humans as species lies at the heart of the Darwinian evolutionary theory and indirectly affects the Victorian political economy. According to Justine Pizzo, Charles Dickens's *Bleak House* is a veritable 'climate model.' His novels speak of the ecological turn, as it shows his concern about fossil fuel energy, implications of waste, entropy, and ecological withering. In the anthropocentric Victorian world, Nature is an antibiotic-resistant microbe and an invasive species. Diseases, unequal distribution of resources, and their understanding of climate emerge together as a 'common global ideology of disaster.' *Dr. Jekyll and Mr. Hyde* presents a parable of a society dependent on the conversion of energy from one form to another. Anne Bronte's ecological metaphors in *Agnes Grey* exhibit her concern for contemporary animal studies and ecocriticism. Conrad's juxtaposition of the city and the forest life presents the adverse effects of imperial civilization and its evil effects on the environment.

In the twentieth century, Darwin destabilized the biological world, Marx the social world, Freud the psychological world, and Nietzsche the metaphorical world. A witness to the two World Wars, T.S. Eliot portrayed the destructive change in the ecological balance by criticizing the negative impacts of modernity. *The Waste Land*, as the title suggests, presents a modern metropolis that is sterile, populated, polluted, contaminated, and supposedly unfit for habitation. The ecological balance structured to the judicious representation of the four elements of Nature – Fire, Air, Earth, and Water – reflects Eliot's concerns for the future of civilization. It is a parable of Nature that highlights human degradation of the contemporary world. The soothing month of April of Chaucer's *Canterbury Tales* has transformed into the 'cruellest month' due to the loss of spirituality and decadence of nature:

> April is the cruellest month, breeding
> Lilacs out of the dead land, mixing
> Memory and desire, stirring
> Dull roots with spring rain.
>
> (102)

The World Wars not only fragmented the human psyche but also alienated man from the natural world. Nature is an inorganic, inanimate object that is described as lifeless, dead land, dull roots, and drought. The apocalypse of Nature symbolizes human depravity which finds expression in *Four Quartets* but has a promise for regeneration at the end. Human exploitation of natural resources leads to discord in the natural world, wherein Eliot searches for physical, psychological, and artistic reconciliation with nature. W.B. Yeats seems to contemplate upon the Romantic idyllic Arcadia and tries to restore innocence and peace in the cold and dry industrialized society. The poet's quest for eternity is reflected through various natural objects of the macrocosm like the archetypal image of a swan in *The Wild Swans at Coole*. He draws extensively from nature, Irish folklore, and mythology to highlight the adverse plight of modern man exposed to the tragedy of collapsing modern civilization due to the brutalities of the World Wars. He seems to use the curative power of art to curb all kinds of disasters both natural and human. Ezra Pound uses unconventional natural metaphors to portray characters and hostile contemporary society. In "The River-Merchant's Wife," the wife's sadness is compared to the sombre monkeys and swiftly spreading moss and in "A Virginal" he compares his lover to a green spring. In the poem "In a Station of the Metro," the faces on a subway platform are compared to the petals on a tree, which testifies to the juxtaposition of modern culture and ecosystem. W.H. Auden's representation of nature is much like the Greeks where the coming of wind denotes glimpses of human temptation and desire, water in streams denote human moods and capacities, landscapes denote human psychological states. The following line from the opening stanza of "Prologue," the introductory poem of Auden's book, *The Orators*, is worth noting:

> By landscape reminded once of his mother's figure
> The mountain heights he remembers get bigger and bigger:

With the finest of mapping pens, he fondly traces
All the family names on the familiar places.

(95)

The present world is thus facing the great wave of Nature's fury which will obliterate us in the yawning future.

III

Ecocriticism has been the slowest of all schools of criticism, emerging as a radical movement and gaining momentum in the 1960s and 1970s. Ecocritics try to find out the environmental damage due to the destructive forms of industrial development and seek for indigenous non-industrial culture to re-establish the alliance between human cultures and the wider environmental movement. They voice their protests for environmental justice by prohibiting acts like the dumping of toxic industrial wastes into water bodies leading to contamination of air, food, and water, loss of lands, and indifference on the part of the government. Robert Pogue Harrison's *Forests: The Shadow of Civilization* highlights the dialectics between nature and civilization. Recent works by ecocritics move beyond the purview of nature writing and romanticism. The poems of Sylvia Plath depict her interest in wild animals, landscape, climate, and pollution as her poems are permeated with images and allusions to nature. Through the various captivating natural images, Plath in "Finisterre" presents the effect of the mist; in "Pheasant," she presents the destruction of nature; in "Full Fathom Five," she presents her father as the sea god; in "Contrusion," she presents the sea as a symbol of terrible loss and loneliness. The following lines from "Wuthering Heights" show how she feels assaulted and lured by the natural world and looks at the ground as a way of searching within:

There is no life higher than the grasstops
Or the hearts of sheep, and the wind
Pours by like destiny bending
Everything in one direction.

(108)

Literature is the best mode to articulate the darkest fear about our future due to the ecological collapse. Linda Hogan's novel *Solar Storms* and Leslie Marmon Silko's novels *Ceremony* and *Almanac of the Dead* focus on highlighting the environmental values in Native American cultures against white industrial capitalism. *American War* by Omar El Akkad, *Annihilation* by Jeff VanderMeer, *Blackfish City* by Sam J. Miller, *A Breath of Fresh Air* by Amulya Malladi, *The Butterfly Effect* by Rajat Chaudhuri, *Death of Grass* by John Christopher, *Hothouse* by Brian W. Aldiss, *Mara and Dann* by Doris Lessing, *Oryx and Crake* and *Surfacing* by Margaret

Atwood are a few literary works which reflect the adverse effects of ecological disasters and climate change aftermath in the contemporary world. To think and analyze a text ecologically, we tend to discover habits of thought of that particular era in the past before the disruption of the human and natural worlds due to the heedless industrialization, embrace of modernism, and the advent of post-modernism. As a solution to the global environmental crisis, we need to awaken from the metropolitan dream and return to our golden past, the pre-lapsarian state of existence.

Down the ages God's supreme creation, Nature, which was once worshipped and nourished as God *per se*, was then tamed my man to use for their own interest. Nature's plight in human history tends to resemble the plight of women in society. Nature is exploited by the human patriarchs to fulfil their libidinal desires. It is tamed and colonized so as to use its resources and yet are not given their due recognition. Every text born within a literary system produces and sustains the culture of its age, reflects the ecophobia of man that is aggravating as we move from a pastoral harmonious world to a metropolitan scientifically modern world.

Thus, we can see that Nature as a subject constantly pervades the realm of English literature across time and space but its mode of representation has metamorphosed with the synchronic change in nature. There is a shift from anthropocentrism to ecocentrism. Eden, the recurrent motif of the Western culture, is now replaced by a lost paradise that needs to be regained through nature narratives that re-embed human beings in the lap of Mother Nature. Ecocriticism demands a re-absorption of a conscious observer of nature and a renewed vision of romantic joy in the contemplation of nature. Literature thus emanates as a truth of ecology rather than a mere ideological screen predicting a projected fantasy or social allegory. It shares a spirit of commitment towards the environmental praxis which is sought after by the ecocritics.

Notes

1 *Natura naturata* is a Latin term coined in the Middle Ages, chiefly used later by Spinoza. It refers to the concept of 'Nature natured' or 'Nature already created.'

2 The Black Death was a bubonic plague pandemic, which reached England in June 1348. The term 'Black Death' was not used until the late seventeenth century.

3 The Peasants' Revolt (1381), also called Wat Tyler's Rebellion, is the first great rebellion in English history. Its immediate cause was the imposition of the poll tax of 1381 which aggravated the dormant economic discontent in parts of England.

4 'Doctrine of Nature' is a theory of natural law beliefs that our civil laws should be based on morality, ethics, and what is inherently correct. This is in contrast to what is called 'positive law' or 'man-made law,' which is defined by statute and common law and may or may not reflect the natural law. Shakespeare's 'Doctrine of Nature' is best reflected in his tragedy of *King Lear*.

5 The French Revolution, which began in May 1789, was a watershed event in modern Europe that shook France between 1787 and 1799. It sought to completely change the relationship between the rulers and the ruled, thereby redefining the concept of the power structure.

Works Cited

Auden, W.H. *The Orator, Selected Poems*. Ed. Edward Mendelson. Vintage International, 1994. 95.

Chaucer, Geoffrey. *The Prologue* to *the Canterbury Tales*. Ed. F.N. Robinson. Oxford University Press, 2009. 73.

Eliot, T.S. *The Waste Land*. Ed. Seamus Perry. Connell Guides, 1994. 102.

Plath, Sylvia. *Wuthering Heights, The Collected Poems*. Ed. Ted Hughes. Harper Perennial, 1992. 108.

Pope, Alexander. *Essay on Criticism*. Ed. Alfred S. West. Cambridge University Press, 1995. 120.

Shakespeare, William. *Julius Caesar*. Ed. David Daniell. Arden Shakespeare, 1998. 184–185.

Wordsworth, William. *Tintern Abbey*. Ed. S.C. Mundhra. Prakash Book Depot, 2004. 115.

For Product Safety Concerns and Information please contact our EU
representative GPSR@taylorandfrancis.com
Taylor & Francis Verlag GmbH, Kaufingerstraße 24, 80331 München, Germany

www.ingramcontent.com/pod-product-compliance
Ingram Content Group UK Ltd.
Pitfield, Milton Keynes, MK11 3LW, UK
UKHW022043061126
477531UK00009B/501
9781032469614